Brief Contents

Part I   Newton's Laws

Chapter 1   Concepts of Motion 2
Chapter 2   Kinematics in One Dimension 33
Chapter 3   Vectors and Coordinate Systems 69
Chapter 4   Kinematics in Two Dimensions 85
Chapter 5   Force and Motion 116
Chapter 6   Dynamics I: Motion Along a Line 138
Chapter 7   Newton's Third Law 167
Chapter 8   Dynamics II: Motion in a Plane 191

Part II   Conservation Laws

Chapter 9   Impulse and Momentum 220
Chapter 10   Energy 245
Chapter 11   Work 278

Part III   Applications of Newtonian Mechanics

Chapter 12   Rotation of a Rigid Body 312
Chapter 13   Newton's Theory of Gravity 354
Chapter 14   Oscillations 377
Chapter 15   Fluids and Elasticity 407

Part IV   Thermodynamics

Chapter 16   A Macroscopic Description of Matter 444
Chapter 17   Work, Heat, and the First Law of Thermodynamics 469
Chapter 18   The Micro/Macro Connection 502
Chapter 19   Heat Engines and Refrigerators 526

Part V   Waves and Optics

Chapter 20   Traveling Waves 560
Chapter 21   Superposition 591
Chapter 22   Wave Optics 627
Chapter 23   Ray Optics 655
Chapter 24   Optical Instruments 694

Part VI   Electricity and Magnetism

Chapter 25   Electric Charges and Forces 720
Chapter 26   The Electric Field 750
Chapter 27   Gauss's Law 780
Chapter 28   The Electric Potential 810
Chapter 29   Potential and Field 839
Chapter 30   Current and Resistance 867
Chapter 31   Fundamentals of Circuits 891
Chapter 32   The Magnetic Field 921
Chapter 33   Electromagnetic Induction 962
Chapter 34   Electromagnetic Fields and Waves 1003
Chapter 35   AC Circuits 1033

Part VII   Relativity and Quantum Physics

Chapter 36   Relativity 1060
Chapter 37   The Foundations of Modern Physics 1102
Chapter 38   Quantization 1125
Chapter 39   Wave Functions and Uncertainty 1156
Chapter 40   One-Dimensional Quantum Mechanics 1179
Chapter 41   Atomic Physics 1216
Chapter 42   Nuclear Physics 1248

Appendix A   Mathematics Review A-1
Appendix B   Periodic Table of Elements A-4
Appendix C   ActivPhysics OnLine Activities and PhET Simulations A-5
Answers to Odd-Numbered Problems A-7

Table of Problem-Solving Strategies

Note for users of the five-volume edition:
Volume 1 (pp. 1–443) includes chapters 1–15.
Volume 2 (pp. 444–559) includes chapters 16–19.
Volume 3 (pp. 560–719) includes chapters 20–24.
Volume 4 (pp. 720–1101) includes chapters 25–36.
Volume 5 (pp. 1102–1279) includes chapters 36–42.

Chapters 37–42 are not in the Standard Edition.

CHAPTER	PROBLEM-SOLVING STRATEGY		PAGE
Chapter 1	**1.1**	**Motion diagrams**	14
Chapter 1	**1.2**	**General problem-solving strategy**	22
Chapter 2	**2.1**	**Kinematics with constant acceleration**	49
Chapter 4	**4.1**	**Projectile motion problems**	94
Chapter 6	**6.1**	**Equilibrium problems**	139
Chapter 6	**6.2**	**Dynamics problems**	142
Chapter 7	**7.1**	**Interacting-objects problems**	175
Chapter 8	**8.1**	**Circular-motion problems**	207
Chapter 9	**9.1**	**Conservation of momentum**	230
Chapter 10	**10.1**	**Conservation of mechanical energy**	255
Chapter 11	**11.1**	**Solving energy problems**	297
Chapter 12	**12.1**	**Rotational dynamics problems**	327
Chapter 12	**12.2**	**Static equilibrium problems**	330
Chapter 17	**17.1**	**Work in ideal-gas processes**	474
Chapter 17	**17.2**	**Calorimetry problems**	484
Chapter 19	**19.1**	**Heat-engine problems**	535
Chapter 21	**21.1**	**Interference of two waves**	613
Chapter 25	**25.1**	**Electrostatic forces and Coulomb's law**	733
Chapter 26	**26.1**	**The electric field of multiple point charges**	752
Chapter 26	**26.2**	**The electric field of a continuous distribution of charge**	758
Chapter 27	**27.1**	**Gauss's law**	795
Chapter 28	**28.1**	**Conservation of energy in charge interactions**	820
Chapter 28	**28.2**	**The electric potential of a continuous distribution of charge**	829
Chapter 31	**31.1**	**Resistor circuits**	906
Chapter 32	**32.1**	**The magnetic field of a current**	928
Chapter 33	**33.1**	**Electromagnetic induction**	976
Chapter 36	**36.1**	**Relativity**	1083
Chapter 40	**40.1**	**Quantum-mechanics problems**	1184

THIRD EDITION

physics

FOR SCIENTISTS AND ENGINEERS
a strategic approach

VOLUME 2

randall d. knight

California Polytechnic State University
San Luis Obispo

PEARSON

Boston Columbus Indianapolis New York San Francisco Upper Saddle River
Amsterdam Cape Town Dubai London Madrid Milan Munich Paris Montreal Toronto
Delhi Mexico City Sao Paulo Sydney Hong Kong Seoul Singapore Taipei Tokyo

Publisher:	James Smith
Senior Development Editor:	Alice Houston, Ph.D.
Senior Project Editor:	Martha Steele
Assistant Editor:	Peter Alston
Media Producer:	Kelly Reed
Senior Administrative Assistant:	Cathy Glenn
Director of Marketing:	Christy Lesko
Executive Marketing Manager:	Kerry McGinnis
Managing Editor:	Corinne Benson
Production Project Manager:	Beth Collins
Production Management, Composition,	
and Interior Design:	Cenveo Publisher Services/Nesbitt Graphics, Inc.
Illustrations:	Rolin Graphics
Cover Design:	Yvo Riezebos Design
Manufacturing Buyer:	Jeff Sargent
Photo Research:	Eric Schrader
Image Lead:	Maya Melenchuk
Cover Printer:	Lehigh-Phoenix
Text Printer and Binder:	R.R. Donnelley/Willard
Cover Image:	Composite illustration by Yvo Riezebos Design
Photo Credits:	See page C-1

Library of Congress Cataloging-in-Publication Data

Knight, Randall Dewey.
Physics for scientists and engineers : a strategic approach / randall d. knight. -- 3rd ed.
    p. cm.
Includes bibliographical references and index.
ISBN 978-0-321-74090-8
1. Physics--Textbooks. I. Title.
QC23.2.K654 2012
530--dc23
2011033849
ISBN-13: 978-0-321-75318-2 ISBN-10: 0-321-75318-6 (Volume 2)
ISBN-13: 978-0-321-74090-8 ISBN-10: 0-321-74090-4 (Student Edition)
ISBN-13: 978-0-321-76519-2 ISBN-10: 0-321-76519-2 (Instructor's Review Copy)
ISBN-13: 978-0-132-83212-0 ISBN-10: 0-132-83212-7 (NASTA Edition)

1 2 3 4 5 6 7 8 9 10—DOW—15 14 13 12 11

www.pearsonhighered.com

About the Author

Randy Knight has taught introductory physics for over 30 years at Ohio State University and California Polytechnic University, where he is currently Professor of Physics. Professor Knight received a bachelor's degree in physics from Washington University in St. Louis and a Ph.D. in physics from the University of California, Berkeley. He was a post-doctoral fellow at the Harvard-Smithsonian Center for Astrophysics before joining the faculty at Ohio State University. It was at Ohio State that he began to learn about the research in physics education that, many years later, led to this book.

Professor Knight's research interests are in the field of lasers and spectroscopy, and he has published over 25 research papers. He also directs the environmental studies program at Cal Poly, where, in addition to introductory physics, he teaches classes on energy, oceanography, and environmental issues. When he's not in the classroom or in front of a computer, you can find Randy hiking, sea kayaking, playing the piano, or spending time with his wife Sally and their seven cats.

Detailed Contents

Volume 1 contains chapters 1-15; Volume 2 contains chapters 16-19; Volume 3 contains chapters 20-24; Volume 4 contains chapters 25-36; Volume 5 contains chapters 36-42.

Part IV Thermodynamics

OVERVIEW   It's All About Energy 443

Chapter 16	A Macroscopic Description of Matter 444
16.1	Solids, Liquids, and Gases 445
16.2	Atoms and Moles 446
16.3	Temperature 449
16.4	Phase Changes 450
16.5	Ideal Gases 452
16.6	Ideal-Gas Processes 456
	SUMMARY 462
	QUESTIONS AND PROBLEMS 463

Chapter 17	Work, Heat, and the First Law of Thermodynamics 469
17.1	It's All About Energy 470
17.2	Work in Ideal-Gas Processes 471
17.3	Heat 475
17.4	The First Law of Thermodynamics 478
17.5	Thermal Properties of Matter 480
17.6	Calorimetry 483

17.7	The Specific Heats of Gases 485
17.8	Heat-Transfer Mechanisms 491
	SUMMARY 495
	QUESTIONS AND PROBLEMS 496

Chapter 18	The Micro/Macro Connection 502
18.1	Molecular Speeds and Collisions 503
18.2	Pressure in a Gas 505
18.3	Temperature 508
18.4	Thermal Energy and Specific Heat 510
18.5	Thermal Interactions and Heat 514
18.6	Irreversible Processes and the Second Law of Thermodynamics 516
	SUMMARY 521
	QUESTIONS AND PROBLEMS 522

Chapter 19	Heat Engines and Refrigerators 526
19.1	Turning Heat into Work 527
19.2	Heat Engines and Refrigerators 529
19.3	Ideal-Gas Heat Engines 534
19.4	Ideal-Gas Refrigerators 538
19.5	The Limits of Efficiency 540
19.6	The Carnot Cycle 542
	SUMMARY 547
	QUESTIONS AND PROBLEMS 548

PART SUMMARY   Thermodynamics 556

Appendix A	Mathematics Review A-1
Appendix B	Periodic Table of Elements A-4
Appendix C	ActivPhysics OnLine Activities and PhET Simulations A-5

Answers to Odd-Numbered Problems A-7

Credits C-1

Index I-1

Thermodynamics

A modern jet engine is a marvel of technical ingenuity. Understanding how a jet engine works requires understanding the thermodynamics of gases and heat engines.

OVERVIEW

It's All About Energy

Thermodynamics—the science of energy in its broadest context—arose hand in hand with the industrial revolution as the systematic study of converting heat energy into mechanical motion and work. Hence the name *thermo + dynamics*. Indeed, the analysis of engines and generators of various kinds remains the focus of engineering thermodynamics. But thermodynamics, as a science, now extends to all forms of energy conversions, including those involving living organisms. For example:

- **Engines** convert the energy of a fuel into the mechanical energy of moving pistons, gears, and wheels.
- **Fuel cells** convert chemical energy into electrical energy.
- **Photovoltaic cells** convert the electromagnetic energy of light into electrical energy.
- **Lasers** convert electrical energy into the electromagnetic energy of light.
- **Organisms** convert the chemical energy of food into a variety of other forms of energy, including kinetic energy, sound energy, and thermal energy.

The major goals of Part IV are to understand both *how* energy transformations such as these take place and *how efficient* they are. We'll discover that the laws of thermodynamics place limits on the efficiency of energy transformations, and understanding these limits is essential for analyzing the very real energy needs of society in the 21st century.

Our ultimate destination in Part IV is an understanding of the thermodynamics of *heat engines*. A heat engine is a device, such as a power plant or an internal combustion engine, that transforms heat energy into useful work. These are the devices that power our modern society.

Understanding how to transform heat into work will be a significant achievement, but we first have many steps to take along the way. We need to understand the concepts of temperature and pressure. We need to learn about the properties of solids, liquids, and gases. Most important, we need to expand our view of energy to include *heat,* the energy that is transferred between two systems at different temperatures.

At a deeper level, we need to see how these concepts are connected to the underlying microphysics of randomly moving molecules. We will find that the familiar concepts of thermodynamics, such as temperature and pressure, have their roots in atomic-level motion and collisions. We will also find it possible to learn a great deal about the properties of molecules, such as their speeds, on the basis of purely macroscopic measurements. This *micro/macro connection* will lead to the second law of thermodynamics, one of the most subtle but also one of the most profound and far-reaching statements in physics.

Only after all these steps have been taken will we be able to analyze a real heat engine. It is an ambitious goal, but one we can achieve.

16 A Macroscopic Description of Matter

The phrase "solid as a rock" is cast in doubt when rocks melt, as they do in this flowing lava.

▶ **Looking Ahead** The goal of Chapter 16 is to learn the characteristics of macroscopic systems.

Temperature

You're familiar with temperature, but what does it actually measure?

We'll start with the simple idea that temperature measures "hotness" and "coldness," but we'll eventually recognize that temperature measures a system's thermal energy.

Before   After

You'll learn to use temperatures in *kelvins*, an absolute temperature scale with absolute zero at 0 K.

Macroscopic Systems

The properties of a macroscopic system are called its **bulk properties.** Examples include a system's volume, density, pressure, and temperature.

Macroscopic systems can be characterized as solid, liquid, or gas—the three most common *phases* of matter.

Starting with this chapter and continuing throughout Part IV, you'll learn to understand the bulk properties of macroscopic systems in terms of the microscopic motion of their atoms and molecules. This **micro/macro connection** is an important part of our modern understanding of matter.

In this chapter, you will learn to describe the amount of substance in a macroscopic system using moles, mass, and the number of atoms.

◀ **Looking Back**
Sections 15.1–15.3 Fluids and pressure

Ideal Gases

We'll model a gas as consisting of tiny, hard spheres that occasionally collide with each other or the walls of their container but otherwise do not interact. You'll learn to use the *ideal-gas law* to understand the bulk properties of a gas.

The Cat's Eye Nebula is a huge ball of hot gas ejected by the star in the center.

Phase Changes

Melting, freezing, boiling, and condensing are *phase changes,* where a macroscopic system changes from one phase to another. You'll learn to represent phase changes in terms of a *phase diagram.*

These melting ice cubes are undergoing a phase change from solid to liquid.

Ideal-Gas Processes

Heating or compressing a gas changes the *state* of the gas. We'll study three basic ideal-gas processes:

■ Constant-pressure process
■ Constant-volume process
■ Constant-temperature process

You'll learn to represent ideal-gas processes on a *pV* diagram.

16.1 Solids, Liquids, and Gases

Each of the elements and most compounds can exist as a solid, liquid, or gas—the three most common **phases** of matter. The change between liquid and solid (freezing or melting) or between liquid and gas (boiling or condensing) is called a **phase change**. We're familiar with only one, or perhaps two, of the phases of most substances because their melting point and/or boiling point are far outside the range of normal human experience. Water is the only substance for which all three phases—ice, liquid, and steam—are everyday occurrences.

NOTE ▶ This use of the word "phase" has no relationship at all to the *phase* or *phase constant* of simple harmonic motion and waves. ◀

Solids, liquids, and gases

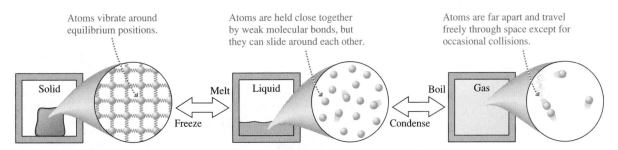

A **solid** is a rigid macroscopic system consisting of particle-like atoms connected by spring-like molecular bonds. Each atom vibrates around an equilibrium position but otherwise has a fixed position. Solids are nearly *incompressible,* which tells us that the atoms in a solid are just about as close together as they can get.

The solid shown here is a **crystal,** meaning that the atoms are arranged in a periodic array. The elements and many compounds have a crystal structure when in their solid phase. In other solids, such as glass, the atoms are frozen into random positions. These are called **amorphous solids.**

A **liquid** is more complicated than either a solid or a gas. Like a solid, a liquid is nearly *incompressible.* This tells us that the molecules in a liquid are about as close together as they can get. Like a gas, a liquid flows and deforms to fit the shape of its container. The fluid nature of a liquid tells us that the molecules are free to move around.

Together, these observations suggest a model in which the molecules of the liquid are loosely held together by weak molecular bonds. The bonds are strong enough that the molecules never get far apart but not strong enough to prevent the molecules from sliding around each other.

A **gas** is a system in which each molecule moves through space as a free, noninteracting particle until, on occasion, it collides with another molecule or with the wall of the container. A gas is a *fluid.* A gas is also highly *compressible,* which tells us that there is lots of space between the molecules.

Gases are fairly simple macroscopic systems; hence many of our examples in Part IV will be based on gases.

State Variables

The parameters used to characterize or describe a macroscopic system are known as **state variables** because, taken all together, they describe the *state* of the macroscopic system. You met some state variables in earlier chapters: volume, pressure, mass, mass density, and thermal energy. We'll soon introduce several new state variables: moles, number density, and, most important, the temperature T.

One important state variable, the mass density, is defined as the ratio of two other state variables:

$$\rho = \frac{M}{V} \quad \text{(mass density)} \quad (16.1)$$

In this chapter we'll use an uppercase M for the system mass and a lowercase m for the mass of an atom. Table 16.1 is a short list of mass densities.

TABLE 16.1 Densities of materials

Substance	ρ (kg/m³)
Air at STP*	1.28
Ethyl alcohol	790
Water (solid)	920
Water (liquid)	1000
Aluminum	2700
Copper	8920
Gold	19,300
Iron	7870
Lead	11,300
Mercury	13,600
Silicon	2330

*$T = 0°C$, $p = 1$ atm

If we change the value of any of the state variables, then we change the state of the system. For example, to *compress* a gas means to decrease its volume. The symbol Δ represents a *change* in the value of a state variable. That is, ΔT is a *change* of temperature and Δp is a *change* of pressure. **For any quantity X, ΔX is always $X_f - X_i$, the final value minus the initial value.**

A system is said to be in **thermal equilibrium** if its state variables are constant and not changing. As an example, a gas is in thermal equilibrium if it has been left undisturbed long enough for p, V, and T to reach steady values.

EXAMPLE 16.1   **The mass of a lead pipe**

A project on which you are working uses a cylindrical lead pipe with outer and inner diameters of 4.0 cm and 3.5 cm, respectively, and a length of 50 cm. What is its mass?

SOLVE The mass density of lead is $\rho_{lead} = 11{,}300 \text{ kg/m}^3$. The volume of a circular cylinder of length l is $V = \pi r^2 l$. In this case we need to find the volume of the outer cylinder, of radius r_2, *minus*

the volume of air in the inner cylinder, of radius r_1. The volume of the pipe is

$$V = \pi r_2^2 l - \pi r_1^2 l = \pi (r_2^2 - r_1^2) l = 1.47 \times 10^{-4} \text{ m}^3$$

Hence the pipe's mass is

$$M = \rho_{lead} V = 1.7 \text{ kg}$$

STOP TO THINK 16.1   The pressure in a system is measured to be 60 kPa. At a later time the pressure is 40 kPa. The value of Δp is

a. 60 kPa   b. 40 kPa   c. 20 kPa   d. −20 kPa

16.2 Atoms and Moles

The mass of a macroscopic system is directly related to the total number of atoms or molecules in the system, denoted N. Because N is determined simply by counting, it is a number with no units. A typical macroscopic system has $N \sim 10^{25}$ atoms, an incredibly large number.

The symbol $\sim$, if you are not familiar with it, stands for "has the order of magnitude." It means that the number is known only to within a factor of 10 or so. The statement $N \sim 10^{25}$, which is read "N is of order 10^{25}," implies that N is somewhere in the range 10^{24} to 10^{26}. It is far less precise than the "approximately equal" symbol $\approx$. Saying $N \sim 10^{25}$ gives us a rough idea of how large N is and allows us to know that it differs significantly from 10^5 or even 10^{15}.

It is often useful to know the number of atoms or molecules per cubic meter in a system. We call this quantity the **number density.** It characterizes how densely the atoms are packed together within the system. In an N-atom system that fills volume V, the number density is

$$\frac{N}{V} \quad \text{(number density)} \qquad (16.2)$$

The SI units of number density are m^{-3}. The number density of atoms in a solid is $(N/V)_{solid} \sim 10^{29} \text{ m}^{-3}$. The number density of a gas depends on the pressure, but is usually less than 10^{27} m^{-3}. As **FIGURE 16.1** shows, **the value of N/V in a *uniform* system is independent of the volume V.** That is, the number density is the same whether you look at the whole system or just a portion of it.

NOTE ▶ While we might say "There are 100 tennis balls per cubic meter," or "There are 10^{29} atoms per cubic meter," tennis balls and atoms are not units. The units of N/V are simply m^{-3}. ◀

FIGURE 16.1 The number density of a uniform system is independent of the volume.

A 100 m³ room has 10,000 tennis balls bouncing around. The number density of tennis balls in the room is $N/V = 10{,}000/100 \text{ m}^3 = 100 \text{ m}^{-3}$.

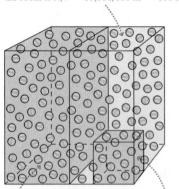

If we look at only half the room, we would find 5000 balls in 50 m³, again giving $N/V = 5000/50 \text{ m}^3 = 100 \text{ m}^{-3}$.

In one-tenth of the room, we would find 1000 balls in 10 m³, again giving $N/V = 1000/10 \text{ m}^3 = 100 \text{ m}^{-3}$.

Atomic Mass and Atomic Mass Number

You will recall from chemistry that atoms of different elements have different masses. The mass of an atom is determined primarily by its most massive constituents, the protons and neutrons in its nucleus. The *sum* of the number of protons and neutrons is called the **atomic mass number** A:

$$A = \text{number of protons} + \text{number of neutrons}$$

A, which by definition is an integer, is written as a leading superscript on the atomic symbol. For example, the common isotope of hydrogen, with one proton and no neutrons, is ^1H. The "heavy hydrogen" isotope called *deuterium*, which includes one neutron, is ^2H. The primary isotope of carbon, with six protons (which makes it carbon) and six neutrons, is ^{12}C. The radioactive isotope ^{14}C, used for carbon dating of archeological finds, contains six protons and eight neutrons.

The **atomic mass** scale is established by defining the mass of ^{12}C to be exactly 12 u, where u is the symbol for the **atomic mass unit.** That is, $m(^{12}\text{C}) = 12$ u. The atomic mass of any other atom is its mass relative to ^{12}C. For example, careful experiments with hydrogen find that the mass *ratio* $m(^1\text{H})/m(^{12}\text{C})$ is 1.0078/12. Thus the atomic mass of hydrogen is $m(^1\text{H}) = 1.0078$ u.

The numerical value of the atomic mass of ^1H is close to, but not exactly, its atomic mass number $A = 1$. For our purposes, it will be sufficient to overlook the slight difference and use the integer atomic mass numbers as the values of the atomic mass. That is, we'll use $m(^1\text{H}) = 1$ u, $m(^4\text{He}) = 4$ u, and $m(^{16}\text{O}) = 16$ u. For molecules, the **molecular mass** is the sum of the atomic masses of the atoms forming the molecule. Thus the molecular mass of O_2, the constituent of oxygen gas, is $m(O_2) = 32$ u.

NOTE ▶ An element's atomic mass number is *not* the same as its atomic number. The *atomic number,* the element's position in the periodic table, is the number of protons in the nucleus. ◄

Table 16.2 shows the atomic mass numbers of some of the elements that we'll use for examples and homework problems. A complete periodic table of the elements, including atomic masses, is found in Appendix B.

One mole of helium, sulfur, copper, and mercury.

TABLE 16.2 Some atomic mass numbers

Element		A
^1H	Hydrogen	1
^4He	Helium	4
^{12}C	Carbon	12
^{14}N	Nitrogen	14
^{16}O	Oxygen	16
^{20}Ne	Neon	20
^{27}Al	Aluminum	27
^{40}Ar	Argon	40
^{207}Pb	Lead	207

Moles and Molar Mass

One way to specify the amount of substance in a macroscopic system is to give its mass. Another is to measure the amount of substance in *moles*. By definition, one **mole** of matter, be it solid, liquid, or gas, is the amount of substance containing as many basic particles as there are atoms in 0.012 kg (12 g) of ^{12}C. Many ingenious experiments have determined that there are 6.02×10^{23} atoms in 0.012 kg of ^{12}C, so we can say that 1 mole of substance, abbreviated 1 mol, is 6.02×10^{23} basic particles.

The basic particle depends on the substance. Helium is a **monatomic gas,** meaning that the basic particle is the helium atom. Thus 6.02×10^{23} helium atoms are 1 mol of helium. But oxygen gas is a **diatomic gas** because the basic particle is the two-atom diatomic molecule O_2. 1 mol of oxygen gas contains 6.02×10^{23} *molecules* of O_2 and thus $2 \times 6.02 \times 10^{23}$ oxygen atoms. Table 16.3 lists the monatomic and diatomic gases that we will use for examples and homework problems.

The number of basic particles per mole of substance is called **Avogadro's number,** N_A. The value of Avogadro's number is

$$N_A = 6.02 \times 10^{23} \text{ mol}^{-1}$$

Despite its name, Avogadro's number is not simply "a number"; it has units. Because there are N_A particles per mole, the number of moles in a substance containing N basic particles is

$$n = \frac{N}{N_A} \quad \text{(moles of substance)} \qquad (16.3)$$

TABLE 16.3 Monatomic and diatomic gases

Monatomic		Diatomic	
He	Helium	H_2	Hydrogen
Ne	Neon	N_2	Nitrogen
Ar	Argon	O_2	Oxygen

Avogadro's number allows us to determine atomic masses in kilograms. Knowing that N_A ^{12}C atoms have a mass of 0.012 kg, the mass of one ^{12}C atom must be

$$m(^{12}\text{C}) = \frac{0.012 \text{ kg}}{6.02 \times 10^{23}} = 1.993 \times 10^{-26} \text{ kg}$$

We defined the atomic mass scale such that $m(^{12}\text{C}) = 12$ u. Thus the conversion factor between atomic mass units and kilograms is

$$1 \text{ u} = \frac{m(^{12}\text{C})}{12} = 1.66 \times 10^{-27} \text{ kg}$$

This conversion factor allows us to calculate the mass in kg of any atom. For example, a ^{20}Ne atom has atomic mass $m(^{20}\text{Ne}) = 20$ u. Multiplying by 1.66×10^{-27} kg/u gives $m(^{20}\text{Ne}) = 3.32 \times 10^{-26}$ kg. If the atomic mass is specified in kilograms, the number of atoms in a system of mass M can be found from

$$N = \frac{M}{m} \tag{16.4}$$

The **molar mass** of a substance is the mass of 1 mol of substance. The molar mass, which we'll designate M_{mol}, has units kg/mol. By definition, the molar mass of ^{12}C is 0.012 kg/mol. For other substances, whose atomic or molecular masses are given relative to ^{12}C, the numerical value of the molar mass is the numerical value of the atomic or molecular mass divided by 1000. For example, the molar mass of He, with $m = 4$ u, is $M_{\text{mol}}(\text{He}) = 0.004$ kg/mol and the molar mass of diatomic O_2 is $M_{\text{mol}}(O_2) = 0.032$ kg/mol.

Equation 16. 4 uses the atomic mass to find the number of atoms in a system. Similarly, you can use the molar mass to determine the number of moles. For a system of mass M consisting of atoms or molecules with molar mass M_{mol},

$$n = \frac{M}{M_{\text{mol}}} \tag{16.5}$$

EXAMPLE 16.2 **Moles of oxygen**

100 g of oxygen gas is how many moles of oxygen?

SOLVE We can do the calculation two ways. First, let's determine the number of molecules in 100 g of oxygen. The diatomic oxygen molecule O_2 has molecular mass $m = 32$ u. Converting this to kg, we get the mass of one molecule:

$$m = 32 \text{ u} \times \frac{1.66 \times 10^{-27} \text{ kg}}{1 \text{ u}} = 5.31 \times 10^{-26} \text{ kg}$$

Thus the number of molecules in 100 g = 0.100 kg is

$$N = \frac{M}{m} = \frac{0.100 \text{ kg}}{5.31 \times 10^{-26} \text{ kg}} = 1.88 \times 10^{24}$$

Knowing the number of molecules gives us the number of moles:

$$n = \frac{N}{N_A} = 3.13 \text{ mol}$$

Alternatively, we can use Equation 16.5 to find

$$n = \frac{M}{M_{\text{mol}}} = \frac{0.100 \text{ kg}}{0.032 \text{ kg/mol}} = 3.13 \text{ mol}$$

STOP TO THINK 16.2 Which system contains more atoms: 5 mol of helium ($A = 4$) or 1 mol of neon ($A = 20$)?

a. Helium. b. Neon. c. They have the same number of atoms.

16.3 Temperature

We are all familiar with the idea of temperature. Mass is a measure of the amount of substance in a system. Velocity is a measure of how fast a system moves. What physical property of the system have you determined if you measure its temperature?

We will begin with the commonsense idea that temperature is a measure of how "hot" or "cold" a system is. As we develop these ideas, we'll find that **temperature** T is related to a system's *thermal energy*. We defined thermal energy in Chapter 10 as the kinetic and potential energy of the atoms and molecules in a system as they vibrate (a solid) or move around (a gas). A system has more thermal energy when it is "hot" than when it is "cold." In Chapter 18, we'll replace these vague notions of hot and cold with a precise relationship between temperature and thermal energy.

To start, we need a means to measure the temperature of a system. This is what a *thermometer* does. A thermometer can be any small macroscopic system that undergoes a measurable change as it exchanges thermal energy with its surroundings. It is placed in contact with a larger system whose temperature it will measure. In a common glass-tube thermometer, for example, a small volume of mercury or alcohol expands or contracts when placed in contact with a "hot" or "cold" object. The object's temperature is determined by the length of the column of liquid.

A thermometer needs a *temperature scale* to be a useful measuring device. In 1742, the Swedish astronomer Anders Celsius sealed mercury into a small capillary tube and observed how it moved up and down the tube as the temperature changed. He selected two temperatures that anyone could reproduce, the freezing and boiling points of pure water, and labeled them 0 and 100. He then marked off the glass tube into one hundred equal intervals between these two reference points. By doing so, he invented the temperature scale that we today call the *Celsius scale*. The units of the Celsius temperature scale are "degrees Celsius," which we abbreviate °C. Note that the degree symbol ° is part of the unit, not part of the number.

The *Fahrenheit scale,* still widely used in the United States, is related to the Celsius scale by

$$T_F = \frac{9}{5}T_C + 32° \tag{16.6}$$

Table 16.4 lists several temperatures measured on the Celsius and Fahrenheit scales and also on the Kelvin scale.

Thermal expansion of the liquid in the thermometer tube pushes it higher in the hot water than in the ice water.

TABLE 16.4 Temperatures measured with different scales

Temperature	$T\,(°C)$	$T\,(K)$	$T\,(°F)$
Melting point of iron	1538	1811	2800
Boiling point of water	100	373	212
Normal body temperature	37	310	99
Room temperature	20	293	68
Freezing point of water	0	273	32
Boiling point of nitrogen	−196	77	−321
Absolute zero	−273	0	−460

Absolute Zero and Absolute Temperature

Any physical property that changes with temperature can be used as a thermometer. In practice, the most useful thermometers have a physical property that changes *linearly* with temperature. One of the most important scientific thermometers is the **constant-volume gas thermometer** shown in FIGURE 16.2a on the next page. This thermometer depends on the fact that the *absolute* pressure (not the gauge pressure) of a gas in a sealed container increases linearly as the temperature increases.

FIGURE 16.2 The pressure in a constant-volume gas thermometer extrapolates to zero at $T_0 = -273°C$. This is the basis for the concept of absolute zero.

(a)

Pressure gauge reading absolute pressure

Rigid gas-filled sphere

System whose temperature is to be measured

T

(b)

Each gas thermometer is calibrated at 0°C and 100°C.

Condensation points

p

Gas 1

Gas 2

Gas 3

T (°C)

-300 -200 -100 0   100

$T_0 = -273°C$

A gas thermometer is first calibrated by recording the pressure at two reference temperatures, such as the boiling and freezing points of water. These two points are plotted on a pressure-versus-temperature graph and a straight line is drawn through them. The gas bulb is then brought into contact with the system whose temperature is to be measured. The pressure is measured, then the corresponding temperature is read off the graph.

FIGURE 16.2b shows the pressure-temperature relationship for three different gases. Notice two important things about this graph.

1. There is a *linear* relationship between temperature and pressure.
2. All gases extrapolate to *zero pressure* at the same temperature: $T_0 = -273°C$. No gas actually gets that cold without condensing, although helium comes very close, but it is surprising that you get the same zero-pressure temperature for any gas and any starting pressure.

The pressure in a gas is due to collisions of the molecules with each other and the walls of the container. A pressure of zero would mean that all motion, and thus all collisions, had ceased. If there were no atomic motion, the system's thermal energy would be zero. The temperature at which all motion would cease, and at which $E_{th} = 0$, is called **absolute zero**. Because temperature is related to thermal energy, absolute zero is the lowest temperature that has physical meaning. We see from the gas-thermometer data that $T_0 = -273°C$.

It is useful to have a temperature scale with the zero point at absolute zero. Such a temperature scale is called an **absolute temperature scale**. Any system whose temperature is measured on an absolute scale will have $T > 0$. The absolute temperature scale having the same unit size as the Celsius scale is called the *Kelvin scale*. It is the SI scale of temperature. The units of the Kelvin scale are *kelvins,* abbreviated as K. The conversion between the Celsius scale and the Kelvin scale is

$$T_K = T_C + 273 \tag{16.7}$$

On the Kelvin scale, absolute zero is 0 K, the freezing point of water is 273 K, and the boiling point of water is 373 K.

NOTE ▶ The units are simply "kelvins," *not* "degrees Kelvin." ◀

FIGURE 16.3 The temperature as a function of time as water is transformed from solid to liquid to gas.

(a)

Ice

Water

(b)

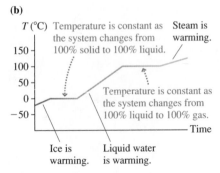

T (°C)   Temperature is constant as the system changes from 100% solid to 100% liquid.

Steam is warming.

150
100
50
0
−50

Temperature is constant as the system changes from 100% liquid to 100% gas.

Time

Ice is warming.

Liquid water is warming.

STOP TO THINK 16.3   The temperature of a glass of water increases from 20°C to 30°C. What is ΔT?

a. 10 K   b. 283 K   c. 293 K   d. 303 K

16.4 Phase Changes

The temperature inside the freezer compartment of a refrigerator is typically about $-20°C$. Suppose you were to remove a few ice cubes from the freezer, place them in a sealed container with a thermometer, then heat them, as FIGURE 16.3a shows. We'll assume that the heating is done so slowly that the inside of the container always has a single, well-defined temperature.

FIGURE 16.3b shows the temperature as a function of time. After steadily rising from the initial $-20°C$, the temperature remains fixed at 0°C for an extended period of time. This is the interval of time during which the ice melts. As it's melting, the ice temperature is 0°C and the liquid water temperature is 0°C. Even though the system is being heated, the liquid water temperature doesn't begin to rise until all the ice has melted. If you were to turn off the flame at any point, the system would remain a mixture of ice and liquid water at 0°C.

NOTE ▶ In everyday language, the three phases of water are called *ice, water,* and *steam.* That is, the term *water* implies the liquid phase. Scientifically, these are the solid, liquid, and gas phases of the compound called *water.* To be clear, we'll use the term *water* in the scientific sense of a collection of H_2O molecules. We'll say either *liquid* or *liquid water* to denote the liquid phase. ◀

The thermal energy of a solid is the kinetic energy of the vibrating atoms plus the potential energy of the stretched and compressed molecular bonds. Melting occurs when the thermal energy gets so large that molecular bonds begin to break, allowing the atoms to move around. The temperature at which a solid becomes a liquid or, if the thermal energy is reduced, a liquid becomes a solid is called the **melting point** or the **freezing point.** Melting and freezing are *phase changes.*

A system at the melting point is in **phase equilibrium,** meaning that any amount of solid can coexist with any amount of liquid. Raise the temperature ever so slightly and the entire system becomes liquid. Lower it slightly and it all becomes solid. But exactly at the melting point the system has no tendency to move one way or the other. That is why the temperature remains constant at the melting point until the phase change is complete.

You can see the same thing happening in Figure 16.3b at 100°C, the boiling point. This is a phase equilibrium between the liquid phase and the gas phase, and any amount of liquid can coexist with any amount of gas at this temperature. Above this temperature, the thermal energy is too large for bonds to be established between molecules, so the system is a gas. If the thermal energy is reduced, the molecules begin to bond with each other and stick together. In other words, the gas condenses into a liquid. The temperature at which a gas becomes a liquid or, if the thermal energy is increased, a liquid becomes a gas is called the **condensation point** or the **boiling point.**

NOTE ▶ Liquid water becomes solid ice at 0°C, but that doesn't mean the temperature of ice is always 0°C. Ice reaches the temperature of its surroundings. If the air temperature in a freezer is −20°C, then the ice temperature is −20°C. Likewise, steam can be heated to temperatures above 100°C. That doesn't happen when you boil water on the stove because the steam escapes, but steam can be heated far above 100°C in a sealed container. ◀

A **phase diagram** is used to show how the phases and phase changes of a substance vary with both temperature and pressure. FIGURE 16.4 shows the phase diagrams for water and carbon dioxide. You can see that each diagram is divided into three regions corresponding to the solid, liquid, and gas phases. The boundary lines separating the regions indicate the phase transitions. The system is in phase equilibrium at a pressure-temperature point that falls on one of these lines.

Phase diagrams contain a great deal of information. Notice on the water phase diagram that the dashed line at $p = 1$ atm crosses the solid-liquid boundary at 0°C and the liquid-gas boundary at 100°C. These well-known melting and boiling point temperatures of water apply only at standard atmospheric pressure. You can see that in Denver, where $p_{atmos} < 1$ atm, water melts at slightly above 0°C and boils at a temperature below 100°C. A *pressure cooker* works by allowing the pressure inside to exceed 1 atm. This raises the boiling point, so foods that are in boiling water are at a temperature above 100°C and cook faster.

Crossing the solid-liquid boundary corresponds to melting or freezing while crossing the liquid-gas boundary corresponds to boiling or condensing. But there's another possibility—crossing the solid-gas boundary. The phase change in which a solid becomes a gas is called **sublimation.** It is not an everyday experience with water, but you probably are familiar with the sublimation of dry ice. Dry ice is solid carbon dioxide. You can see on the carbon dioxide phase diagram that the dashed line at $p = 1$ atm crosses the solid-*gas* boundary, rather than the solid-liquid boundary, at $T = -78$°C. This is the *sublimation temperature* of dry ice.

FIGURE 16.4 Phase diagrams (not to scale) for water and carbon dioxide.

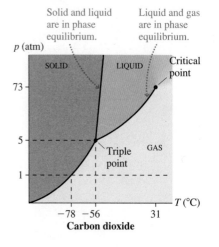

Liquid carbon dioxide does exist, but only at pressures greater than 5 atm and temperatures greater than $-56°C$. A CO_2 fire extinguisher contains *liquid* carbon dioxide under high pressure. (You can hear the liquid slosh if you shake a CO_2 fire extinguisher.)

One important difference between the water and carbon dioxide phase diagrams is the slope of the solid-liquid boundary. For most substances, the solid phase is denser than the liquid phase and the liquid is denser than the gas. Pressurizing the substance compresses it and increases the density. If you start compressing CO_2 gas at room temperature, thus moving upward through the phase diagram along a vertical line, you'll first condense it to a liquid and eventually, if you keep compressing, change it into a solid.

Water is a very unusual substance in that the density of ice is *less* than the density of liquid water. That is why ice floats. If you compress ice, making it denser, you eventually cause a phase transition in which the ice turns to liquid water! Consequently, the solid-liquid boundary for water slopes to the left.

The liquid-gas boundary ends at a point called the **critical point.** Below the critical point, liquid and gas are clearly distinct and there is a phase change if you go from one to the other. But there is no clear distinction between liquid and gas at pressures or temperatures above the critical point. The system is a *fluid,* but it can be varied continuously between high density and low density without a phase change.

The final point of interest on the phase diagram is the **triple point** where the phase boundaries meet. Two phases are in phase equilibrium along the boundaries. The triple point is the *one* value of temperature and pressure for which all three phases can coexist in phase equilibrium. That is, any amounts of solid, liquid, and gas can happily coexist at the triple point. For water, the triple point occurs at $T_3 = 0.01°C$ and $p_3 = 0.006$ atm.

The significance of the triple point of water is its connection to the Kelvin temperature scale. The Celsius scale required two *reference points,* the boiling and melting points of water. We can now see that these are not very satisfactory reference points because their values vary as the pressure changes. In contrast, there's only one temperature at which ice, liquid water, and water vapor will coexist in equilibrium. If you produce this equilibrium in the laboratory, then you *know* the system is at the triple-point temperature.

The triple-point temperature of water is an ideal reference point, hence the Kelvin temperature scale is *defined* to be a linear temperature scale starting from 0 K at absolute zero and passing through 273.16 K at the triple point of water. Because $T_3 = 0.01°C$, absolute zero on the Celsius scale is $T_0 = -273.15°C$.

NOTE ▶ To be consistent with our use of significant figures, $T_0 = -273$ K is the appropriate value to use in calculations *unless* you know other temperatures with an accuracy of better than $1°C$. ◀

Food takes longer to cook at high altitudes because the boiling point of water is less than 100°C.

STOP TO THINK 16.4 For which is there a sublimation temperature that is higher than a melting temperature?

a. Water   b. Carbon dioxide   c. Both   d. Neither

16.5 Ideal Gases

We noted earlier in the chapter that solids and liquids are nearly incompressible, an observation suggesting that atoms are fairly hard and cannot be pressed together once they come into contact with each other. Based on this observation, suppose we were to model atoms as "hard spheres" that do not interact except for occasional elastic collisions when two atoms come into contact and bounce apart.

This is a *model* of an atom—what we might call the *ideal atom*—because it ignores the weak attractive interactions that hold liquids and solids together. A gas of these noninteracting atoms is called an **ideal gas.** It is a gas of small, hard, randomly moving atoms that bounce off each other and the walls of their container but otherwise do not interact. The ideal gas is a somewhat simplified description of a real gas, but experiments show that the ideal-gas model is quite good for real gases if two conditions are met:

1. The density is low (i.e., the atoms occupy a volume much smaller than that of the container), and
2. The temperature is well above the condensation point.

If the density gets too high, or the temperature too low, then the attractive forces between the atoms begin to play an important role and our model, which ignores those attractive forces, fails. These are the forces that are responsible, under the right conditions, for the gas condensing into a liquid.

We've been using the term "atoms," but many gases, as you know, consist of molecules rather than atoms. Only helium, neon, argon, and the other inert elements in the far-right column of the periodic table of the elements form monatomic gases. Hydrogen (H_2), nitrogen (N_2), and oxygen (O_2) are diatomic gases. As far as translational motion is concerned, the ideal-gas model does not distinguish between a monatomic gas and a diatomic gas; both are considered as simply small, hard spheres. Hence the terms "atoms" and "molecules" can be used interchangeably to mean the basic constituents of the gas.

The Ideal-Gas Law

Section 16.1 introduced the idea of *state variables,* those parameters that describe the state of a macroscopic system. The state variables for an ideal gas are the volume V of its container, the number of moles n of the gas present in the container, the temperature T of the gas and its container, and the pressure p that the gas exerts on the walls of the container. These four state parameters are not independent of each other. If you change the value of one—by, say, raising the temperature—then one or more of the others will change as well. Each change of the parameters is a *change of state* of the system.

Experiments during the 17th and 18th centuries found a very specific relationship between the four state variables. Suppose you change the state of a gas, by heating it or compressing it or doing something else to it, and measure p, V, n, and T. Repeat this many times, changing the state of the gas each time, until you have a large table of p, V, n, and T values.

Then make a graph on which you plot pV, the product of the pressure and volume, on the vertical axis and nT, the product of the number of moles and temperature (in kelvins), on the horizontal axis. The very surprising result is that for *any* gas, whether it is hydrogen or helium or oxygen or methane, **you get exactly the same graph,** the linear graph shown in FIGURE 16.5. In other words, nothing about the graph indicates what gas was used because all gases give the same result.

NOTE ▶ No real gas could extend to $nT = 0$ because it would condense. But an ideal gas never condenses because the only interactions among the molecules are hard-sphere collisions. ◀

As you can see, there is a very clear proportionality between the quantity pV and the quantity nT. If we designate the slope of the line in this graph as R, then we can write the relationship as

$$pV = R \times (nT)$$

It is customary to write this relationship in a slightly different form, namely

$$pV = nRT \qquad \text{(ideal-gas law)} \qquad (16.8)$$

Equation 16.8 is the **ideal-gas law. The ideal-gas law is a relationship among the four state variables—p, V, n, and T—that characterize a gas in thermal equilibrium.**

FIGURE 16.5 A graph of pV versus nT for an ideal gas.

The constant R, which is determined experimentally as the slope of the graph in Figure 16.5, is called the **universal gas constant.** Its value, in SI units, is

$$R = 8.31 \text{ J/mol K}$$

The units of R seem puzzling. The denominator mol K is clear because R multiplies nT. But what about the joules? The left side of the ideal-gas law, pV, has units

$$\text{Pa m}^3 = \frac{\text{N}}{\text{m}^2}\text{m}^3 = \text{N m} = \text{joules}$$

The product pV has units of joules, as shown on the vertical axis in Figure 16.5.

NOTE ▶ You perhaps learned in chemistry to work gas problems using units of atmospheres and liters. To do so, you had a different numerical value of R expressed in those units. In physics, however, we always work gas problems in SI units. Pressures *must* be in Pa, volumes in m^3, and temperatures in K. ◀

The surprising fact, and one worth commenting upon, is that *all* gases have the *same* graph and the *same* value of R. There is no obvious reason a very simple atomic gas such as helium should have the same slope as a more complex gas such as methane (CH_4). Nonetheless, both turn out to have the same value for R. The ideal-gas law, within its limits of validity, describes *all* gases with a single value of the constant R.

EXAMPLE 16.3 **Calculating a gas pressure**

100 g of oxygen gas is distilled into an evacuated 600 cm^3 container. What is the gas pressure at a temperature of 150°C?

MODEL The gas can be treated as an ideal gas. Oxygen is a diatomic gas of O_2 molecules.

SOLVE From the ideal-gas law, the pressure is $p = nRT/V$. In Example 16.2 we calculated the number of moles in 100 g of O_2 and found $n = 3.13$ mol. Gas problems typically involve several conversions to get quantities into the proper units, and this example is no exception. The SI units of V and T are m^3 and K, respectively, thus

$$V = (600 \text{ cm}^3)\left(\frac{1 \text{ m}}{100 \text{ cm}}\right)^3 = 6.00 \times 10^{-4}\text{ m}^3$$

$$T = (150 + 273) \text{ K} = 423 \text{ K}$$

With this information, the pressure is

$$p = \frac{nRT}{V} = \frac{(3.13 \text{ mol})(8.31 \text{ J/mol K})(423 \text{ K})}{6.00 \times 10^{-4}\text{ m}^3}$$

$$= 1.83 \times 10^7 \text{ Pa} = 181 \text{ atm}$$

In this text we will consider only gases in sealed containers. The number of moles (and number of molecules) will not change during a problem. In that case,

$$\frac{pV}{T} = nR = \text{constant} \tag{16.9}$$

If the gas is initially in state i, characterized by the state variables p_i, V_i, and T_i, and at some later time in a final state f, the state variables for these two states are related by

$$\frac{p_f V_f}{T_f} = \frac{p_i V_i}{T_i} \quad \text{(ideal gas in a sealed container)} \tag{16.10}$$

This before-and-after relationship between the two states, reminiscent of a conservation law, will be valuable for many problems.

EXAMPLE 16.4 **Calculating a gas temperature**

A cylinder of gas is at 0°C. A piston compresses the gas to half its original volume and three times its original pressure. What is the final gas temperature?

MODEL Treat the gas as an ideal gas in a sealed container.

SOLVE The before-and-after relationship of Equation 16.10 can be written

$$T_2 = T_1 \frac{p_2}{p_1}\frac{V_2}{V_1}$$

In this problem, the compression of the gas results in $V_2/V_1 = \frac{1}{2}$ and $p_2/p_1 = 3$. The initial temperature is $T_1 = 0°C = 273$ K. With this information,

$$T_2 = 273 \text{ K} \times 3 \times \frac{1}{2} = 409 \text{ K} = 136°C$$

ASSESS We did not need to know actual values of the pressure and volume, just the *ratios* by which they change.

We will often want to refer to the number of molecules N in a gas rather than the number of moles n. This is an easy change to make. Because $n = N/N_A$, the ideal-gas law in terms of N is

$$pV = nRT = \frac{N}{N_A}RT = N\frac{R}{N_A}T \qquad (16.11)$$

R/N_A, the ratio of two known constants, is known as **Boltzmann's constant** k_B:

$$k_B = \frac{R}{N_A} = 1.38 \times 10^{-23} \text{ J/K}$$

The subscript B distinguishes Boltzmann's constant from a spring constant or other uses of the symbol k.

Ludwig Boltzmann was an Austrian physicist who did some of the pioneering work in statistical physics during the mid-19th century. Boltzmann's constant k_B can be thought of as the "gas constant per molecule," whereas R is the "gas constant per mole." With this definition, the ideal-gas law in terms of N is

$$pV = Nk_BT \qquad \text{(ideal-gas law)} \qquad (16.12)$$

Equations 16.8 and 16.12 are both the ideal-gas law, just expressed in terms of different state variables.

Recall that the number density (molecules per m^3) was defined as N/V. A rearrangement of Equation 16.12 gives the number density as

$$\frac{N}{V} = \frac{p}{k_BT} \qquad (16.13)$$

This is a useful consequence of the ideal-gas law, but keep in mind that the pressure *must* be in SI units of pascals and the temperature *must* be in SI units of kelvins.

EXAMPLE 16.5 **The distance between molecules**

"Standard temperature and pressure," abbreviated **STP,** are $T = 0°C$ and $p = 1$ atm. Estimate the average distance between gas molecules at STP.

MODEL Consider the gas to be an ideal gas.

SOLVE Suppose a container of volume V holds N molecules at STP. How do we estimate the distance between them? Imagine placing an imaginary sphere around each molecule, separating it from its neighbors. This divides the total volume V into N little spheres of volume v_i, where $i = 1$ to N. The spheres of two neighboring molecules touch each other, like a crate full of Ping-Pong balls of somewhat different sizes all touching their neighbors, so the distance between two molecules is the sum of the radii of their two spheres. Each of these spheres is somewhat different, but a reasonable *estimate* of the distance between molecules would be twice the *average* radius of a sphere.

The average volume of one of these little spheres is

$$v_{avg} = \frac{V}{N} = \frac{1}{N/V}$$

That is, the average volume per molecule (m^3 per molecule) is the inverse of the number density, the number of molecules per m^3. This is not the volume of the molecule itself, which is much smaller, but the average volume of space that each molecule can claim as its own. We can use Equation 16.13 to calculate the number density:

$$\frac{N}{V} = \frac{p}{k_BT} = \frac{1.01 \times 10^5 \text{ Pa}}{(1.38 \times 10^{-23} \text{ J/K})(273 \text{ K})}$$

$$= 2.69 \times 10^{25} \text{ molecules/m}^3$$

where we used the definition of STP in SI units. Thus the average volume per molecule is

$$v_{avg} = \frac{1}{N/V} = 3.72 \times 10^{-26} \text{ m}^3$$

The volume of a sphere is $\frac{4}{3}\pi r^3$, so the average radius of a sphere is

$$r_{avg} = \left(\frac{3}{4\pi}v_{avg}\right)^{1/3} = 2.1 \times 10^{-9} \text{ m} = 2.1 \text{ nm}$$

The average distance between two molecules, with their spheres touching, is twice r_{avg}. Thus

$$\text{average distance} = 2r_{avg} \approx 4 \text{ nm}$$

This is a simple estimate, so we've given the answer with only one significant figure.

ASSESS One of the assumptions of the ideal-gas model is that atoms or molecules are "far apart" in comparison to the sizes of atoms and molecules. Chemistry experiments find that small molecules, such as N_2 and O_2, are roughly 0.3 nm in diameter. For a gas at STP, we see that the average distance between molecules is more than 10 times the size of a molecule. Thus the ideal-gas model works very well for a gas at STP.

You have two containers of equal volume. One is full of helium gas. The other holds an equal mass of nitrogen gas. Both gases have the same pressure. How does the temperature of the helium compare to the temperature of the nitrogen?

a. $T_{helium} > T_{nitrogen}$ b. $T_{helium} = T_{nitrogen}$ c. $T_{helium} < T_{nitrogen}$

16.6 Ideal-Gas Processes

The ideal-gas law is the connection between the state variables pressure, temperature, and volume. If the state variables change, as they would from heating or compressing the gas, the state of the gas changes. An **ideal-gas process** is the means by which the gas changes from one state to another.

NOTE ▶ Even in a sealed container, the ideal-gas law is a relationship among *three* variables. In general, *all three change* during an ideal-gas process. As a result, thinking about cause and effect can be rather tricky. Don't make the mistake of thinking that one variable is constant unless you're sure, beyond a doubt, that it is. ◀

The *pV* Diagram

It will be very useful to represent ideal-gas processes on a graph called a **pV diagram.** This is nothing more than a graph of pressure versus volume. The important idea behind the *pV* diagram is that *each point* on the graph represents a single, unique state of the gas. That seems surprising at first, because a point on the graph only directly specifies the values of *p* and *V*. But knowing *p* and *V*, and assuming that *n* is known for a sealed container, we can find the temperature by using the ideal-gas law. Thus each point actually represents a triplet of values (*p, V, T*) specifying the state of the gas.

For example, FIGURE 16.6a is a *pV* diagram showing three states of a system consisting of 1 mol of gas. The values of *p* and *V* can be read from the axes for each point, then the temperature at that point determined from the ideal-gas law. An ideal-gas process can be represented as a "trajectory" in the *pV* diagram. The trajectory shows all the intermediate states through which the gas passes. FIGURE 16.6b shows two different processes by which the gas of Figure 16.6a can be changed from state 1 to state 3.

There are infinitely many ways to change the gas from state 1 to state 3. Although the initial and final states are the same for each of them, the particular process by which the gas changes—that is, the particular trajectory—will turn out to have very real consequences. For example, you will soon learn that the work done by an expanding gas, a quantity of very practical importance in various devices, depends on the trajectory followed. The *pV* diagram is an important graphical representation of the process.

FIGURE 16.6 The state of the gas and ideal-gas processes can be shown on a *pV* diagram.

(a)   Each state of an ideal gas is represented as a point on a *pV* diagram.

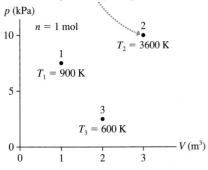

(b)   Two different processes that change the gas from state 1 to state 3.

Quasi-Static Processes

Strictly speaking, the ideal-gas law applies only to gases in *thermal equilibrium,* meaning that the state variables are constant and not changing. But, by definition, an ideal-gas process causes some of the state variables to change. The gas is *not* in thermal equilibrium while the process of changing from state 1 to state 2 is under way.

To use the ideal-gas law throughout, we will assume that the process occurs *so slowly* that the system is never far from equilibrium. In other words, the values of *p*, *V*, and *T* at any point in the process are essentially the same as the equilibrium values they would assume if we stopped the process at that point. A process that is essentially in thermal equilibrium at all times is called a **quasi-static process.** It is an idealization, like a frictionless surface, but one that is a very good approximation in many real situations.

An important characteristic of a quasi-static process is that the trajectory through the *pV* diagram can be *reversed*. If you quasi-statically expand a gas by slowly pulling a piston out, as shown in FIGURE 16.7a, you can reverse the process by slowly pushing the piston in. The gas retraces its *pV* trajectory until it has returned to its initial state. Contrast this with what happens when the membrane bursts in FIGURE 16.7b. That is a sudden process, not at all quasi-static. The *irreversible* process of Figure 16.7b cannot be represented on a *pV* diagram.

The critical question is: How slow must a process be to qualify as quasi-static? That is a difficult question to answer. This textbook will always assume that processes are quasi-static. It turns out to be a reasonable assumption for the types of examples and homework problems we will look at. Irreversible processes will be left to more advanced courses.

Constant-Volume Process

Many important gas processes take place in a container of constant, unchanging volume. A constant-volume process is called an **isochoric process,** where *iso* is a prefix meaning "constant" or "equal" while *choric* is from a Greek root meaning "volume." An isochoric process is one for which

$$V_f = V_i \tag{16.14}$$

For example, suppose that you have a gas in the closed, rigid container shown in FIGURE 16.8a. Warming the gas with a Bunsen burner will raise its pressure without changing its volume. This process is shown as the vertical line $1 \rightarrow 2$ on the *pV* diagram of FIGURE 16.8b. A constant-volume cooling, by placing the container on a block of ice, would lower the pressure and be represented as the vertical line from 2 to 1. **Any isochoric process appears on a *pV* diagram as a vertical line.**

FIGURE 16.7 The slow motion of the piston is a quasi-static process. The bursting of the membrane is not.

(a)
Piston moves very slowly.

Quasi-static process

(b)
Membrane suddenly bursts.

Irreversible process

FIGURE 16.8 A constant-volume (isochoric) process.

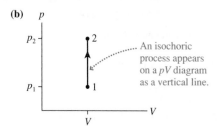

EXAMPLE 16.6 **A constant-volume gas thermometer**

A constant-volume gas thermometer is placed in contact with a reference cell containing water at the triple point. After reaching equilibrium, the gas pressure is recorded as 55.78 kPa. The thermometer is then placed in contact with a sample of unknown temperature. After the thermometer reaches a new equilibrium, the gas pressure is 65.12 kPa. What is the temperature of this sample?

MODEL The thermometer's volume doesn't change, so this is an isochoric process.

SOLVE The temperature at the triple point of water is $T_1 = 0.01°C = 273.16$ K. The ideal-gas law for a closed system

is $p_2V_2/T_2 = p_1V_1/T_1$. The volume doesn't change, so $V_2/V_1 = 1$. Thus

$$T_2 = T_1 \frac{V_2}{V_1} \frac{p_2}{p_1} = T_1 \frac{p_2}{p_1} = (273.16 \text{ K}) \frac{65.12 \text{ kPa}}{55.78 \text{ kPa}}$$

$$= 318.90 \text{ K} = 45.75°C$$

The temperature *must* be in kelvins to do this calculation, although it is common to convert the final answer to °C. The fact that the pressures were given to four significant figures justified using $T_K = T_C + 273.15$ rather than the usual $T_C + 273$.

ASSESS $T_2 > T_1$, which we expected from the increase in pressure.

FIGURE 16.9 A constant-pressure (isobaric) process.

(a) The piston's mass maintains a constant pressure in the cylinder.

Before   After

(b)

Piston

(c) *p* An isobaric process appears on a *pV* diagram as a horizontal line.

Constant-Pressure Process

Other gas processes take place at a constant, unchanging pressure. A constant-pressure process is called an **isobaric process,** where *baric* is from the same root as "barometer" and means "pressure." An isobaric process is one for which

$$p_f = p_i \qquad (16.15)$$

FIGURE 16.9a shows one method of changing the state of a gas while keeping the pressure constant. A cylinder of gas has a tight-fitting piston of mass *M* that can slide up and down but seals the container so that no atoms enter or escape. As the free-body diagram of FIGURE 16.9b shows, the piston and the air press down with force $p_{atmos}A + Mg$ while the gas inside pushes up with force $p_{gas}A$. In equilibrium, the gas pressure inside the cylinder is

$$p_{gas} = p_{atmos} + \frac{Mg}{A} \qquad (16.16)$$

In other words, the gas pressure is determined by the requirement that the gas must support both the mass of the piston and the air pressing inward. **This pressure is independent of the temperature of the gas or the height of the piston, so it stays constant as long as *M* is unchanged.**

If the cylinder is warmed, the gas will expand and push the piston up. But the pressure, determined by mass *M*, will not change. This process is shown on the *pV* diagram of FIGURE 16.9c as the horizontal line 1 → 2. We call this an *isobaric expansion.* An *isobaric compression* occurs if the gas is cooled, lowering the piston. **Any isobaric process appears on a *pV* diagram as a horizontal line.**

EXAMPLE 16.7 **Comparing pressure**

The two cylinders in FIGURE 16.10 contain ideal gases at 20°C. Each cylinder is sealed by a frictionless piston of mass *M*.

a. How does the pressure of gas 2 compare to that of gas 1? Is it larger, smaller, or the same?

b. Suppose gas 2 is warmed to 80°C. Describe what happens to the pressure and volume.

FIGURE 16.10 Compare the pressures of the two gases.

MODEL Treat the gases as ideal gases.

SOLVE a. The pressure in the gas is determined by the requirement that the piston be in mechanical equilibrium. The pressure of the gas inside pushes up on the piston; the air pressure and the weight of the piston press down. The gas pressure $p = p_{atmos} + Mg/A$ depends on the mass of the piston, but not at all on how high the piston is or what type of gas is inside the cylinder. Thus both pressures are the same.

b. Neither does the pressure depend on temperature. Warming the gas increases the temperature, but the pressure—determined by the mass and area of the piston—is unchanged. Because $pV/T = $ constant, and *p* is constant, it must be true that $V/T = $ constant. As *T* increases, the volume *V* also must increase to keep V/T unchanged. In other words, increasing the gas temperature causes the volume to expand—the piston goes up—but with no change in pressure. This is an isobaric process.

EXAMPLE 16.8 **Identifying a gas**

Your lab assistant distilled 50 g of a gas into a cylinder, but he left without writing down what kind of gas it is. The cylinder has a pressure regulator that adjusts a piston to keep the pressure at a constant 2.00 atm. To identify the gas, you measure the cylinder volume at several different temperatures, acquiring the data shown at the right. What is the gas?

Temperature (°C)	Volume (L)
−50	11.6
0	14.0
50	16.2
100	19.4
150	21.8

MODEL The pressure doesn't change, so heating the gas is an iso-baric process.

SOLVE The ideal-gas law is $pV = nRT$. Writing this as

$$V = \frac{nR}{p}T$$

we see that a graph of V versus T should be a straight line passing through the origin. Further, we can use the slope of the graph, nR/p, to measure the number of moles of gas, and from that we can identify the gas by determining its molar mass.

FIGURE 16.11 is a graph of the data, with the volumes and temperatures converted to SI units of m^3 (1 m^3 = 1000 L) and kelvins. The y-intercept of the graph is essentially zero, confirming the behavior of the gas as ideal, and the slope of the best-fit line is 5.16×10^{-5} m^3/K. The number of moles of gas is

$$n = \frac{p}{R} \times \text{slope} = \frac{2 \times 101,300 \text{ Pa}}{8.31 \text{ J/mol K}} \times 5.16 \times 10^{-5} \text{ m}^3/K = 1.26 \text{ mol}$$

From this, the molar mass is

FIGURE 16.11 A graph of the gas volume versus its temperature.

$$M = \frac{0.050 \text{ kg}}{1.26 \text{ mol}} = 0.040 \text{ kg/mol}$$

Thus the atomic mass is 40 u, identifying the gas as argon.

ASSESS The atomic mass is that of a well-known gas, which gives us confidence in the result.

STOP TO THINK 16.6 Two cylinders of equal diameter contain the same number of moles of the same ideal gas. Each cylinder is sealed by a frictionless piston. To have the same pressure in both cylinders, which piston would you use in cylinder 2?

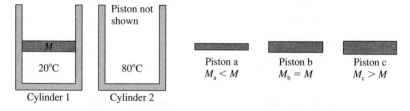

Constant-Temperature Process

The last process we wish to look at for now is one that takes place at a constant temperature. A constant-temperature process is called an **isothermal process**. An isothermal process is one for which $T_f = T_i$. Because $pV = nRT$, a constant-temperature process in a closed system (constant n) is one for which the product pV doesn't change. Thus

$$p_f V_f = p_i V_i \tag{16.17}$$

in an isothermal process.

One possible isothermal process is illustrated in **FIGURE 16.12a**, where a piston is being pushed down to compress a gas. If the piston is pushed *slowly,* then heat energy transfer through the walls of the cylinder keeps the gas at the same temperature as the surrounding liquid. This is an *isothermal compression*. The reverse process, with the piston slowly pulled out, would be an *isothermal expansion*.

Representing an isothermal process on the pV diagram is a little more complicated than the two previous processes because both p and V change. As long as T remains fixed, we have the relationship

$$p = \frac{nRT}{V} = \frac{\text{constant}}{V} \tag{16.18}$$

The inverse relationship between p and V causes the graph of an isothermal process to be a *hyperbola*. As one state variable goes up, the other goes down.

FIGURE 16.12 A constant-temperature (isothermal) process.

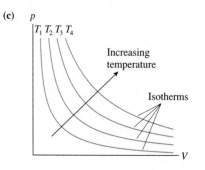

The process shown as $1 \rightarrow 2$ in FIGURE 16.12b represents the *isothermal compression* shown in Figure 16.12a. An *isothermal expansion* would move in the opposite direction along the hyperbola.

The location of the hyperbola depends on the value of T. A lower-temperature process is represented by a hyperbola closer to the origin than a higher-temperature process. FIGURE 16.12c shows four hyperbolas representing the temperatures T_1 to T_4, where $T_4 > T_3 > T_2 > T_1$. These are called **isotherms.** A gas undergoing an isothermal process moves along the isotherm of the appropriate temperature.

EXAMPLE 16.9 **Compressing air in the lungs**

An ocean snorkeler takes a deep breath at the surface, filling his lungs with 4.0 L of air. He then descends to a depth of 5.0 m. At this depth, what is the volume of air in the snorkeler's lungs?

MODEL At the surface, the pressure in the lungs is 1.00 atm. Because the body cannot sustain large pressure differences between inside and outside, the air pressure in the lungs rises—and the volume decreases—to match the surrounding water pressure as he descends.

SOLVE The ideal-gas law for a sealed container is

$$V_2 = \frac{p_1}{p_2}\frac{T_2}{T_1}V_1$$

Air is quickly warmed to body temperature as it enters through the nose and mouth, and it remains at body temperature as the snorkeler dives, so $T_2/T_1 = 1$. We know $p_1 = 1.00 \text{ atm} = 101,300 \text{ Pa}$ at the surface. We can find p_2 from the hydrostatic pressure equation, using the density of seawater:

$$p_2 = p_1 + \rho g d = 101,300 \text{ Pa} + (1030 \text{ kg/m}^3)(9.80 \text{ m/s}^2)(5.0 \text{ m})$$

$$= 151,800 \text{ Pa}$$

With this, the volume of the lungs at a depth of 5.0 m is

$$V_2 = \frac{101,300 \text{ Pa}}{151,800 \text{ Pa}} \times 1 \times 4.0 \text{ L} = 2.7 \text{ L}$$

ASSESS The air inside your lungs does compress—significantly—as you dive below the surface.

EXAMPLE 16.10 **A multistep process**

A gas at 2.0 atm pressure and a temperature of 200°C is first expanded isothermally until its volume has doubled. It then undergoes an isobaric compression until it returns to its original volume. First show this process on a pV diagram. Then find the final temperature (in °C) and pressure.

MODEL The final state of the isothermal expansion is the initial state for an isobaric compression.

VISUALIZE FIGURE 16.13 shows the process. As the gas expands isothermally, it moves downward along an isotherm until it reaches volume $V_2 = 2V_1$. The gas is then compressed at constant pressure p_2 until its final volume V_3 equals its original volume V_1. State 3 is on an isotherm closer to the origin, so we expect to find $T_3 < T_1$.

SOLVE $T_2/T_1 = 1$ during the isothermal expansion and $V_2 = 2V_1$, so the pressure at point 2 is

$$p_2 = p_1 \frac{T_2}{T_1}\frac{V_1}{V_2} = p_1 \frac{V_1}{2V_1} = \frac{1}{2}p_1 = 1.0 \text{ atm}$$

We have $p_3/p_2 = 1$ during the isobaric compression and $V_3 = V_1 = \frac{1}{2}V_2$, so

FIGURE 16.13 A pV diagram for the process of Example 16.10.

$$T_3 = T_2 \frac{p_3}{p_2}\frac{V_3}{V_2} = T_2 \frac{\frac{1}{2}V_2}{V_2} = \frac{1}{2}T_2 = 236.5 \text{ K} = -36.5°C$$

where we converted T_2 to 473 K before doing calculations and then converted T_3 back to °C. The final state, with $T_3 = -36.5°C$ and $p_3 = 1.0 \text{ atm}$, is one in which both the pressure and the absolute temperature are half their original values.

STOP TO THINK 16.7   What is the ratio T_f/T_i for this process?

a. $\frac{1}{4}$
b. $\frac{1}{2}$
c. 1 (no change)
d. 2
e. 4
f. There's not enough information to tell.

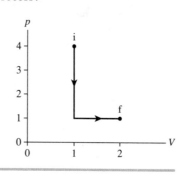

CHALLENGE EXAMPLE 16.11   **Depressing a piston**

A large, 50.0-cm-diameter metal cylinder filled with air supports a 20.0 kg piston that can slide up and down without friction. The piston is 100.0 cm above the bottom when the temperature is 20°C. An 80.0 kg student then stands on the piston. After several minutes have elapsed, by how much has the piston been depressed?

MODEL The metal walls of the cylinder are a good thermal conductor, so after several minutes the gas temperature—even if it initially changed—will return to room temperature. The final temperature matches the initial temperature. Assume that the atmospheric pressure is 1 atm.

VISUALIZE FIGURE 16.14 shows the cylinder before and after the student stands on it. The volume of the cylinder is $V = Ah$, and only h changes.

SOLVE The ideal-gas law for a sealed container is

$$\frac{p_2 A h_2}{T_2} = \frac{p_1 A h_1}{T_1}$$

Because $T_2 = T_1$, the final height of the piston is

$$h_2 = \frac{p_1}{p_2} h_1$$

The pressure of the gas is determined by the mass of the piston (and anything on the piston) and the pressure of the air above. In equilibrium,

$$p = p_{atmos} + \frac{Mg}{A} = \begin{cases} 1.023 \times 10^5 \text{ Pa} & \text{piston only} \\ 1.063 \times 10^5 \text{ Pa} & \text{piston and student} \end{cases}$$

where we used $p_{atmos} = 1$ atm $= 1.013 \times 10^5$ Pa and $A = \pi r^2 = 0.196$ m². The final height of the piston is

FIGURE 16.14 The student compresses the gas.

$$h_2 = \frac{1.023 \times 10^5 \text{ Pa}}{1.063 \times 10^5 \text{ Pa}} \times 100.0 \text{ cm} = 96.2 \text{ cm}$$

The question, however, was by how much the piston is depressed. This is $h_1 - h_2 = 3.8$ cm.

ASSESS Neither the piston nor the student increases the gas pressure to much above 1 atm, so it's not surprising that the added weight of the student doesn't push the piston down very far.

SUMMARY

The goal of Chapter 16 has been to learn the characteristics of macroscopic systems.

General Principles

Three Common Phases of Matter

Solid   Rigid, definite shape. Nearly incompressible.

Liquid   Molecules loosely held together by molecular bonds, but able to move around. Nearly incompressible.

Gas   Molecules moving freely through space. Compressible.

The different phases exist for different conditions of temperature T and pressure p. The boundaries separating the regions of a phase diagram are lines of phase equilibrium. Any amounts of the two phases can coexist in equilibrium. The **triple point** is the one value of temperature and pressure at which all three phases can coexist in equilibrium.

Important Concepts

Ideal-Gas Model

- Atoms and molecules are small, hard spheres that travel freely through space except for occasional collisions with each other or the walls.

- The model is valid when the density is low and the temperature well above the condensation point.

Ideal-Gas Law

The **state variables** of an ideal gas are related by the ideal-gas law

$$pV = nRT \quad \text{or} \quad pV = Nk_BT$$

where $R = 8.31$ J/mol K is the universal gas constant and $k_B = 1.38 \times 10^{-23}$ J/K is Boltzmann's constant. p, V, and T *must* be in SI units of Pa, m^3, and K.

For a gas in a sealed container, with constant n:

$$\frac{p_2 V_2}{T_2} = \frac{p_1 V_1}{T_1}$$

Counting atoms and moles

A macroscopic sample of matter consists of N atoms (or molecules), each of mass m (the **atomic** or **molecular mass**):

$$N = \frac{M}{m}$$

Alternatively, we can state that the sample consists of n **moles:**

$$n = \frac{N}{N_A} \quad \text{or} \quad \frac{M}{M_{mol}}$$

where $N_A = 6.02 \times 10^{23}$ mol^{-1} is **Avogadro's number.**

The molar mass M_{mol}, in kg/mol, is the numerical value of the atomic or molecular mass in u divided by 1000. The atomic or molecular mass, in atomic mass units u, is well approximated by the **atomic mass number** A. The atomic mass unit is

$$1 \text{ u} = 1.66 \times 10^{-27} \text{ kg}$$

The **number density** of the sample is $\frac{N}{V}$.

Applications

Temperature scales

$$T_F = \frac{9}{5}T_C + 32° \qquad T_K = T_C + 273$$

The Kelvin temperature scale is based on:

- Absolute zero at $T_0 = 0$ K
- The triple point of water at $T_3 = 273.16$ K

Three basic gas processes

1. **Isochoric,** or constant volume
2. **Isobaric,** or constant pressure
3. **Isothermal,** or constant temperature

Terms and Notation

bulk properties	atomic mass number, A	absolute zero, T_0	ideal-gas law
micro/macro connection	atomic mass	absolute temperature scale	universal gas constant, R
phase	atomic mass unit, u	melting point	Boltzmann's constant, k_B
phase change	molecular mass	freezing point	STP
solid	mole, n	phase equilibrium	ideal-gas process
crystal	monatomic gas	condensation point	pV diagram
amorphous solid	diatomic gas	boiling point	quasi-static process
liquid	Avogadro's number, N_A	phase diagram	isochoric process
gas	molar mass, M_{mol}	sublimation	isobaric process
state variable	temperature, T	critical point	isothermal process
thermal equilibrium	constant-volume gas	triple point	isotherm
number density, N/V	thermometer	ideal gas	

CONCEPTUAL QUESTIONS

1. Rank in order, from highest to lowest, the temperatures $T_1 = 0$ K, $T_2 = 0°C$, and $T_3 = 0°F$.

2. The sample in an experiment is initially at 10°C. If the sample's temperature is doubled, what is the new temperature in °C?

3. a. Is there a highest temperature at which ice can exist? If so, what is it? If not, why not?
   b. Is there a lowest temperature at which water vapor can exist? If so, what is it? If not, why not?

4. The cylinder in FIGURE Q16.4 is divided into two compartments by a frictionless piston that can slide back and forth. Is the pressure on the left side greater than, less than, or equal to the pressure on the right? Explain.

FIGURE Q16.4

5. A gas is in a sealed container. By what factor does the gas temperature change if:
   a. The volume is doubled and the pressure is tripled?
   b. The volume is halved and the pressure is tripled?

6. A gas is in a sealed container. The gas pressure is tripled and the temperature is doubled.
   a. What happens to the number of moles of gas in the container?
   b. What happens to the number density of the gas in the container?

7. An aquanaut lives in an underwater apartment 100 m beneath the surface of the ocean. Compare the freezing and boiling points of water in the aquanaut's apartment to their values at the surface. Are they higher, lower, or the same? Explain.

8. a. A sample of water vapor in an enclosed cylinder has an initial pressure of 500 Pa at an initial temperature of −0.01°C. A piston squeezes the sample smaller and smaller, without limit. Describe what happens to the water as the squeezing progresses.
   b. Repeat part a if the initial temperature is 0.03°C warmer.

9. A gas is in a sealed container. By what factor does the gas pressure change if:
   a. The volume is doubled and the temperature is tripled?
   b. The volume is halved and the temperature is tripled?

10. A gas undergoes the process shown in FIGURE Q16.10. By what factor does the temperature change?

FIGURE Q16.10   FIGURE Q16.11

11. The temperature increases from 300 K to 1200 K as a gas undergoes the process shown in FIGURE Q16.11. What is the final pressure?

12. A student is asked to sketch a pV diagram for a gas that goes through a cycle consisting of (a) an isobaric expansion, (b) a constant-volume reduction in temperature, and (c) an isothermal process that returns the gas to its initial state. The student draws the diagram shown in FIGURE Q16.12. What, if anything, is wrong with the student's diagram?

FIGURE Q16.12

EXERCISES AND PROBLEMS

Problems labeled ▦ integrate material from earlier chapters.

Exercises

Section 16.1 Solids, Liquids, and Gases

1. | What volume of water has the same mass as 100 cm³ of gold?
2. ‖ The nucleus of a uranium atom has a diameter of 1.5×10^{-14} m and a mass of 4.0×10^{-25} kg. What is the density of the nucleus?
3. ‖ A hollow aluminum sphere with outer diameter 10.0 cm has a mass of 690 g. What is the sphere's inner diameter?
4. ‖ What is the diameter of a copper sphere that has the same mass as a 10 cm × 10 cm × 10 cm cube of aluminum?

Section 16.2 Atoms and Moles

5. ‖ How many atoms are in a 2.0 cm × 2.0 cm × 2.0 cm cube of aluminum?
6. ‖ How many moles are in a 2.0 cm × 2.0 cm × 2.0 cm cube of copper?
7. ‖ What is the number density of (a) aluminum and (b) lead?
8. ‖ An element in its solid phase has mass density 1750 kg/m³ and number density 4.39×10^{28} atoms/m³. What is the element's atomic mass number?
9. ‖ What volume of aluminum has the same number of atoms as 10 cm³ of mercury?
10. ‖ 1.0 mol of gold is shaped into a sphere. What is the sphere's diameter?

Section 16.3 Temperature

Section 16.4 Phase Changes

11. | The lowest and highest natural temperatures ever recorded on earth are −127°F in Antarctica and 136°F in Libya. What are these temperatures in °C and in K?
12. | At what temperature does the numerical value in °F match the numerical value in °C?
13. ‖ A demented scientist creates a new temperature scale, the "Z scale." He decides to call the boiling point of nitrogen 0°Z and the melting point of iron 1000°Z.
    a. What is the boiling point of water on the Z scale?
    b. Convert 500°Z to degrees Celsius and to kelvins.
14. ‖ What is the temperature in °F and the pressure in Pa at the triple point of (a) water and (b) carbon dioxide?

Section 16.5 Ideal Gases

15. | A cylinder contains nitrogen gas. A piston compresses the gas to half its initial volume. Afterward,
    a. Has the mass density of the gas changed? If so, by what factor? If not, why not?
    b. Has the number of moles of gas changed? If so, by what factor? If not, why not?
16. ‖ 3.0 mol of gas at a temperature of −120°C fills a 2.0 L container. What is the gas pressure?
17. ‖ A gas at 100°C fills volume V_0. If the pressure is held constant, what is the volume if (a) the Celsius temperature is doubled and (b) the Kelvin temperature is doubled?

18. ‖ A rigid container holds 2.0 mol of gas at a pressure of 1.0 atm and a temperature of 30°C.
    a. What is the container's volume?
    b. What is the pressure if the temperature is raised to 130°C?
19. ‖ The total lung capacity of a typical adult is 5.0 L. Approxi-
    BIO mately 20% of the air is oxygen. At sea level and at a body temperature of 37°C, how many oxygen molecules do the lungs contain at the end of a strong inhalation?
20. ‖ A 20-cm-diameter cylinder that is 40 cm long contains 50 g of oxygen gas at 20°C.
    a. How many moles of oxygen are in the cylinder?
    b. How many oxygen molecules are in the cylinder?
    c. What is the number density of the oxygen?
    d. What is the reading of a pressure gauge attached to the tank?
21. ‖ A 10-cm-diameter cylinder of neon gas is 30 cm long and at 30°C. The pressure gauge reads 120 psi. What is the mass density of the gas?

Section 16.6 Ideal-Gas Processes

22. | A gas with initial state variables p_1, V_1, and T_1 expands isothermally until $V_2 = 2V_1$. What are (a) T_2 and (b) p_2?
23. | A gas with initial state variables p_1, V_1, and T_1 is cooled in an isochoric process until $p_2 = \frac{1}{3}p_1$. What are (a) V_2 and (b) T_2?
24. | A rigid sphere is submerged in boiling water in a room where the air pressure is 1.0 atm. The sphere has an open valve with its inlet just above the water level. After a long period of time has elapsed, the valve is closed. What will be the pressure inside the sphere if it is then placed in (a) a mixture of ice and water and (b) an insulated box filled with dry ice?
25. ‖ A rigid container holds hydrogen gas at a pressure of 3.0 atm and a temperature of 20°C. What will the pressure be if the temperature is lowered to −20°C?
26. ‖ A 24-cm-diameter vertical cylinder is sealed at the top by a frictionless 20 kg piston. The piston is 84 cm above the bottom when the gas temperature is 303°C. The air above the piston is at 1.00 atm pressure.
    a. What is the gas pressure inside the cylinder?
    b. What will the height of the piston be if the temperature is lowered to 15°C?
27. ‖ 0.10 mol of argon gas is admitted to an evacuated 50 cm³ container at 20°C. The gas then undergoes an isochoric heating to a temperature of 300°C.
    a. What is the final pressure of the gas?
    b. Show the process on a pV diagram. Include a proper scale on both axes.
28. | 0.10 mol of argon gas is admitted to an evacuated 50 cm³ container at 20°C. The gas then undergoes an isobaric heating to a temperature of 300°C.
    a. What is the final volume of the gas?
    b. Show the process on a pV diagram. Include a proper scale on both axes.
29. ‖ 0.10 mol of argon gas is admitted to an evacuated 50 cm³ container at 20°C. The gas then undergoes an isothermal expansion to a volume of 200 cm³.
    a. What is the final pressure of the gas?
    b. Show the process on a pV diagram. Include a proper scale on both axes.

30. | 0.0040 mol of gas undergoes the process shown in FIG-URE EX16.30.
    a. What type of process is this?
    b. What are the initial and final temperatures in °C?

FIGURE EX16.30   FIGURE EX16.31

31. ‖ A gas with an initial temperature of 900°C undergoes the process shown in FIGURE EX16.31.
    a. What type of process is this?
    b. What is the final temperature in °C?
    c. How many moles of gas are there?

32. | 0.020 mol of gas undergoes the process shown in FIG-URE EX16.32.
    a. What type of process is this?
    b. What is the final temperature in °C?
    c. What is the final volume V_2?

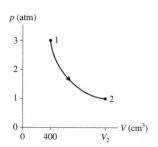

FIGURE EX16.32

Problems

33. ‖ The atomic mass number of copper is $A = 64$. Assume that atoms in solid copper form a cubic crystal lattice. To envision this, imagine that you place atoms at the centers of tiny sugar cubes, then stack the little sugar cubes to form a big cube. If you dissolve the sugar, the atoms left behind are in a cubic crystal lattice. What is the smallest distance between two copper atoms?

34. ‖ An element in its solid phase forms a cubic crystal lattice (see Problem 33) with mass density 7950 kg/m³. The smallest spacing between two adjacent atoms is 0.227 nm. What is the element's atomic mass number?

35. ‖ The molecular mass of water (H_2O) is $A = 18$. How many protons are there in 1.0 L of liquid water?

36. ‖ Estimate the number density of gas molecules in the earth's atmosphere at sea level.

37. | The solar corona is a very hot atmosphere surrounding the visible surface of the sun. X-ray emissions from the corona show that its temperature is about 2×10^6 K. The gas pressure in the corona is about 0.03 Pa. Estimate the number density of particles in the solar corona.

38. ‖ The semiconductor industry manufactures integrated circuits in large vacuum chambers where the pressure is 1.0×10^{-10} mm of Hg.
    a. What fraction is this of atmospheric pressure?
    b. At $T = 20°C$, how many molecules are in a cylindrical chamber 40 cm in diameter and 30 cm tall?

39. ‖ A 6.0-cm-diameter, 10-cm-long cylinder contains 100 mg of oxygen (O_2) at a pressure less than 1 atm. The cap on one end of the cylinder is held in place only by the pressure of the air. One day when the atmospheric pressure is 100 kPa, it takes a 184 N force to pull the cap off. What is the temperature of the gas?

40. ‖ A nebula—a region of the galaxy where new stars are forming—contains a very tenuous gas with 100 atoms/cm³. This gas is heated to 7500 K by ultraviolet radiation from nearby stars. What is the gas pressure in atm?

41. ‖‖ An inflated bicycle inner tube is 2.2 cm in diameter and 200 cm in circumference. A small leak causes the gauge pressure to decrease from 110 psi to 80 psi on a day when the temperature is 20°C. What mass of air is lost? Assume the air is pure nitrogen.

42. ‖ On average, each person in the industrialized world is responsible for the emission of 10,000 kg of carbon dioxide (CO_2) every year. This includes CO_2 that you generate directly, by burning fossil fuels to operate your car or your furnace, as well as CO_2 generated on your behalf by electric generating stations and manufacturing plants. CO_2 is a greenhouse gas that contributes to global warming. If you were to store your yearly CO_2 emissions in a cube at STP, how long would each edge of the cube be?

43. ‖ A gas at temperature T_0 and atmospheric pressure fills a cylinder. The gas is transferred to a new cylinder with three times the volume, after which the pressure is half the original pressure. What is the new temperature of the gas?

44. ‖ To determine the mass of neon contained in a rigid, 2.0 L cylinder, you vary the cylinder's temperature while recording the reading of a pressure gauge. Your data are as follows:

Temperature (°C)	Pressure gauge (atm)
100	6.52
150	7.80
200	8.83
250	9.59

Use the best-fit line of an appropriate graph to determine the mass of the neon.

45. ‖ The 3.0-m-long pipe in FIGURE P16.45 is closed at the top end. It is slowly pushed straight down into the water until the top end of the pipe is level with the water's surface. What is the length L of the trapped volume of air?

FIGURE P16.45

46. ‖ An electric generating plant boils water to produce high-pressure steam. The steam spins a turbine that is connected to the generator.
    a. How many liters of water must be boiled to fill a 5.0 m^3 boiler with 50 atm of steam at 400°C?
    b. The steam has dropped to 2.0 atm pressure at 150°C as it exits the turbine. How much volume does it now occupy?

47. ‖‖ On a cool morning, when the temperature is 15°C, you measure the pressure in your car tires to be 30 psi. After driving 20 mi on the freeway, the temperature of your tires is 45°C. What pressure will your tire gauge now show?

48. ‖ The air temperature and pressure in a laboratory are 20°C and 1.0 atm. A 1.0 L container is open to the air. The container is then sealed and placed in a bath of boiling water. After reaching thermal equilibrium, the container is opened. How many moles of air escape?

49. ‖ A gas cylinder with a tight-fitting, movable piston contains 200 cm^3 of air at 1.0 atm. It floats on the surface of a swimming pool filled with 15°C water. The cylinder is then pulled slowly underwater to a depth of 3.0 m. What is the volume of gas at this depth?

50. ‖ The mercury manometer shown in FIGURE P16.50 is attached to a gas cell. The mercury height h is 120 mm when the cell is placed in an ice-water mixture. The mercury height drops to 30 mm when the device is carried into an industrial freezer. What is the freezer temperature?
    Hint: The right tube of the manometer is much narrower than the left tube. What reasonable assumption can you make about the gas volume?

FIGURE P16.50   FIGURE P16.51

51. ‖ The U-shaped tube in FIGURE P16.51 has a total length of 1.0 m. It is open at one end, closed at the other, and is initially filled with air at 20°C and 1.0 atm pressure. Mercury is poured slowly into the open end without letting any air escape, thus compressing the air. This is continued until the open side of the tube is completely filled with mercury. What is the length L of the column of mercury?

52. ‖ A diver 50 m deep in 10°C fresh water exhales a 1.0-cm-diameter bubble. What is the bubble's diameter just as it reaches the surface of the lake, where the water temperature is 20°C?
    Hint: Assume that the air bubble is always in thermal equilibrium with the surrounding water.

53. ‖ A compressed-air cylinder is known to fail if the pressure exceeds 110 atm. A cylinder that was filled to 25 atm at 20°C is stored in a warehouse. Unfortunately, the warehouse catches fire and the temperature reaches 950°C. Does the cylinder blow?

54. ‖ Reproduce FIGURE P16.54 on a piece of paper. A gas starts with pressure p_1 and volume V_1. Show on the figure the process in which the gas undergoes an isochoric process that doubles the pressure, then an isobaric process that doubles the volume, followed by an isothermal process that doubles the volume again. Label each of the three processes.

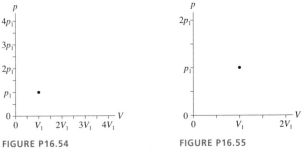

FIGURE P16.54   FIGURE P16.55

55. ‖ Reproduce FIGURE P16.55 on a piece of paper. A gas starts with pressure p_1 and volume V_1. Show on the figure the process in which the gas undergoes an isothermal process during which the volume is halved, then an isochoric process during which the pressure is halved, followed by an isobaric process during which the volume is doubled. Label each of the three processes.

56. ‖ 8.0 g of helium gas follows the process $1 \rightarrow 2 \rightarrow 3$ shown in FIGURE P16.56. Find the values of V_1, V_3, p_2, and T_3.

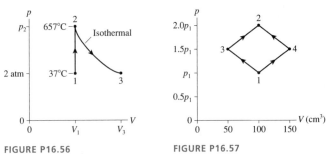

FIGURE P16.56   FIGURE P16.57

57. ‖ FIGURE P16.57 shows two different processes by which 1.0 g of nitrogen gas moves from state 1 to state 2. The temperature of state 1 is 25°C. What are (a) pressure p_1 and (b) temperatures (in °C) T_2, T_3, and T_4?

58. ‖ FIGURE P16.58 shows two different processes by which 80 mol of gas move from state 1 to state 2. The dashed line is an isotherm.
    a. What is the temperature of the isothermal process?
    b. What maximum temperature is reached along the straight-line process?

FIGURE P16.58   FIGURE P16.59

59. ‖ 0.10 mol of gas undergoes the process $1 \rightarrow 2$ shown in FIGURE P16.59.
    a. What are temperatures T_1 and T_2 (in °C)?
    b. What type of process is this?
    c. The gas undergoes an isothermal compression from point 2 until the volume is restored to the value it had at point 1. What is the final pressure of the gas?

60. | 0.0050 mol of gas undergoes the process $1 \rightarrow 2 \rightarrow 3$ shown in FIGURE P16.60. What are (a) temperature T_1, (b) pressure p_2, and (c) volume V_3?

FIGURE P16.60

FIGURE P16.61

61. ‖ 4.0 g of oxygen gas, starting at 20°C, follow the process $1 \rightarrow 2$ shown in FIGURE P16.61. What is temperature T_2 (in °C)?
62. ‖ 10 g of dry ice (solid CO_2) is placed in a 10,000 cm³ container, then all the air is quickly pumped out and the container sealed. The container is warmed to 0°C, a temperature at which CO_2 is a gas.
    a. What is the gas pressure? Give your answer in atm.
    The gas then undergoes an isothermal compression until the pressure is 3.0 atm, immediately followed by an isobaric compression until the volume is 1000 cm³.
    b. What is the final temperature of the gas (in °C)?
    c. Show the process on a pV diagram.
63. ‖ A container of gas at 2.0 atm pressure and 127°C is compressed at constant temperature until the volume is halved. It is then further compressed at constant pressure until the volume is halved again.
    a. What are the final pressure and temperature of the gas?
    b. Show this process on a pV diagram.
64. ‖ Five grams of nitrogen gas at an initial pressure of 3.0 atm and at 20°C undergo an isobaric expansion until the volume has tripled.
    a. What is the gas volume after the expansion?
    b. What is the gas temperature after the expansion (in °C)?
    The gas pressure is then decreased at constant volume until the original temperature is reached.
    c. What is the gas pressure after the decrease?
    Finally, the gas is isothermally compressed until it returns to its initial volume.
    d. What is the final gas pressure?
    e. Show the full three-step process on a pV diagram. Use appropriate scales on both axes.

In Problems 65 through 68 you are given the equation(s) used to solve a problem. For each of these, you are to
    a. Write a realistic problem for which this is the correct equation(s).
    b. Draw a pV diagram.
    c. Finish the solution of the problem.

65. $p_2 = \dfrac{300 \text{ cm}^3}{100 \text{ cm}^3} \times 1 \times 2 \text{ atm}$

66. $(T_2 + 273) \text{ K} = \dfrac{200 \text{ kPa}}{500 \text{ kPa}} \times 1 \times (400 + 273) \text{ K}$

67. $V_2 = \dfrac{(400 + 273) \text{ K}}{(50 + 273) \text{ K}} \times 1 \times 200 \text{ cm}^3$

68. $(2.0 \times 101{,}300 \text{ Pa})(100 \times 10^{-6} \text{ m}^3) = n(8.31 \text{ J/mol K})T_1$

$n = \dfrac{0.12 \text{ g}}{20 \text{ g/mol}}$

$T_2 = \dfrac{200 \text{ cm}^3}{100 \text{ cm}^3} \times 1 \times T_1$

Challenge Problems

69. The 50 kg lead piston shown in FIGURE CP16.69 floats on 0.12 mol of compressed air.
    a. What is the piston height h if the temperature is 30°C?
    b. How far does the piston move if the temperature is increased by 100°C?

FIGURE CP16.69

70. A diving bell is a 3.0-m-tall cylinder closed at the upper end but open at the lower end. The temperature of the air in the bell is 20°C. The bell is lowered into the ocean until its lower end is 100 m deep. The temperature at that depth is 10°C.
    a. How high does the water rise in the bell after enough time has passed for the air inside to reach thermal equilibrium?
    b. A compressed-air hose from the surface is used to expel all the water from the bell. What minimum air pressure is needed to do this?
71. 10,000 cm³ of 200°C steam at a pressure of 20 atm is cooled until it condenses. What is the volume of the liquid water? Give your answer in cm³.
72. The cylinder in FIGURE CP16.72 has a moveable piston attached to a spring. The cylinder's cross-section area is 10 cm², it contains 0.0040 mol of gas, and the spring constant is 1500 N/m. At 20°C the spring is neither compressed nor stretched. How far is the spring compressed if the gas temperature is raised to 100°C?

FIGURE CP16.72

73. Containers A and B in FIGURE CP16.73 hold the same gas. The volume of B is four times the volume of A. The two containers are connected by a thin tube (negligible volume) and a valve that is closed. The gas in A is at 300 K and pressure of 1.0×10^5 Pa. The gas in B is at 400 K and pressure of 5.0×10^5 Pa. Heaters will maintain the temperatures of A and B even after the valve is opened.
    a. After the valve is opened, gas will flow one way or the other until A and B have equal pressure. What is this final pressure?
    b. Is this a reversible or an irreversible process? Explain.

FIGURE CP16.73

74. The closed cylinder of FIGURE CP16.74 has a tight-fitting but fric-tionless piston of mass M. The piston is in equilibrium when the left chamber has pressure p_0 and length L_0 while the spring on the right is compressed by ΔL.
    a. What is ΔL in terms of p_0, L_0, A, M, and k?
    b. Suppose the piston is moved a small distance x to the right. Find an expression for the net force $(F_x)_{net}$ on the piston. As-sume all motions are slow enough for the gas to remain at the same temperature as its surroundings.
    c. If released, the piston will oscillate around the equilibrium position. Assuming $x \ll L_0$ find an expression for the oscil-lation period T.

Hint: Use the binomial approximation.

FIGURE CP16.74

STOP TO THINK ANSWERS

Stop to Think 16.1: d. The pressure *decreases* by 20 kPa.

Stop to Think 16.2: a. The number of atoms depends only on the number of moles, not the substance.

Stop to Think 16.3: a. The step size on the Kelvin scale is the same as the step size on the Celsius scale. A *change* of 10°C is a *change* of 10 K.

Stop to Think 16.4: a. On the water phase diagram, you can see that for a pressure just slightly below the triple-point pressure, the solid/gas transition occurs at a higher temperature than does the solid/liquid transition at high pressures. This is not true for carbon dioxide.

Stop to Think 16.5: c. $T = pV/nR$. Pressure and volume are the same, but n differs. The number of moles in mass M is $n = M/M_{mol}$. Helium, with the smaller molar mass, has a larger number of moles and thus a lower temperature.

Stop to Think 16.6: b. The pressure is determined entirely by the weight of the piston pressing down. Changing the temperature chang-es the volume of the gas, but not its pressure.

Stop to Think 16.7: b. The temperature decreases by a factor of 4 during the isochoric process, where $p_f/p_i = \frac{1}{4}$. The temperature then increases by a factor of 2 during the isobaric expansion, where $V_f/V_i = 2$.

17 Work, Heat, and the First Law of Thermodynamics

This false-color thermal image—an infrared photo—shows where heat is escaping from the house.

▶ **Looking Ahead** The goal of Chapter 17 is to develop and apply the first law of thermodynamics.

Energy Transfers

There are two ways to transfer energy between a system and its environment: work and heat.

Work is the transfer of energy in a *mechanical interaction*—when external forces push or pull on the system.

You will learn to calculate the work done in an ideal-gas process as the negative of the area under a pV curve.

Heat is the transfer of energy in a *thermal interaction*—when the system and its environment have different temperatures.

You will learn to calculate the heat energy required for various processes. For example, it takes just over 4000 J of heat to raise the temperature of 1 kg of water by 1°C.

◀ **Looking Back**
Section 11.4 Work

The First Law of Thermodynamics

The first law of thermodynamics is a very general statement of the idea that energy can be transferred and transformed but not created or destroyed.

- The system's thermal energy increases if energy is transferred into the system as heat or work.
- The system's thermal energy decreases if energy is transferred out of the system as heat or work.

In a jet engine, part of the heat from burning fuel is used to do work—pushing the aircraft forward. The remainder becomes increased thermal energy of the hot exhaust gases. No energy is destroyed.

◀ **Looking Back**
Sections 11.7–11.8 Conservation of energy

Thermal Properties of Matter

Changing the thermal energy can cause

- A temperature change, or
- A phase change.

A material's response to heat is governed by its *specific heat*, its *heat of fusion*, and its *heat of vaporization*.

You'll learn how to do practical calorimetry calculations to determine the final temperature of two or more interacting systems.

◀ **Looking Back**
Sections 16.4–16.6 Phase changes and ideal gases

Heat Transfer

Heat energy can be transferred between a system and its environment by

- Conduction
- Radiation
- Convection
- Evaporation

Air heated by the hot teakettle streams upward, an example of heat transfer by convection.

17.1 It's All About Energy

A key idea of Chapter 11 was the work-kinetic energy theorem in the form

$$\Delta K = W_c + W_{diss} + W_{ext} \tag{17.1}$$

Equation 17.1 tells us that the kinetic energy of a system of particles is changed when forces do work on the particles by pushing or pulling them through a distance. Here

1. W_c is the work done by conservative forces. This work can be represented as a change in the system's potential energy: $\Delta U = -W_c$.
2. W_{diss} is the work done by friction-like dissipative forces within the system. This work increases the system's thermal energy: $\Delta E_{th} = -W_{diss}$.
3. W_{ext} is the work done by external forces that originate in the environment. The push of a piston rod would be an external force.

With these definitions, Equation 17.1 becomes

$$\Delta K + \Delta U + \Delta E_{th} = W_{ext} \tag{17.2}$$

The system's *mechanical energy* was defined as $E_{mech} = K + U$. FIGURE 17.1 reminds you that the mechanical energy is associated with the motion of the system as a whole, while E_{th} is associated with the motion of the atoms and molecules within the system. E_{mech} is the *macroscopic* energy of the system as a whole while E_{th} is the *microscopic* energy of the particle-like atoms and spring-like molecular bonds. This led to our final energy statement of Chapter 11:

$$\Delta E_{sys} = \Delta E_{mech} + \Delta E_{th} = W_{ext} \tag{17.3}$$

Thus the total energy of an *isolated system,* for which $W_{ext} = 0$, is constant. This was the essence of the law of conservation of energy as stated in Chapter 11.

The emphasis in Chapters 10 and 11 was on isolated systems. There we were interested in learning how kinetic and potential energy were *transformed* into each other and, where there is friction, into thermal energy. Now we want to focus on how energy is *transferred* between the system and its environment, when W_{ext} is *not* zero.

NOTE ▶ Strictly speaking, Equation 17.3 should use the *internal energy* E_{int} rather than the thermal energy E_{th}, where $E_{int} = E_{th} + E_{chem} + E_{nuc} + \cdots$ includes all the various kinds of energies that can be stored inside a system. This textbook will focus on simple thermodynamics systems in which the internal energy is entirely thermal: $E_{int} = E_{th}$. We'll leave other forms of internal energy to more advanced courses. ◀

Energy Transfer

Doing work on a system can have very different consequences. FIGURE 17.2a shows an object being lifted at steady speed by a rope. The rope's tension is an external force doing work W_{ext} on the system. In this case, the energy transferred into the system goes entirely to increasing the system's macroscopic potential energy U_{grav}, part of the mechanical energy. The energy-transfer process $W_{ext} \rightarrow E_{mech}$ is shown graphically in the energy bar chart of Figure 17.2a.

Contrast this with FIGURE 17.2b, where the same rope with the same tension now drags the object at steady speed across a rough surface. The tension does the same amount of work, but the mechanical energy does not change. Instead, friction increases the thermal energy of the object + surface system. The energy-transfer process $W_{ext} \rightarrow E_{th}$ is shown in the energy bar chart of Figure 17.2b.

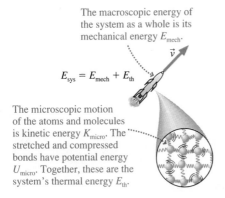

FIGURE 17.1 The total energy of a system consists of the macroscopic energy plus the microscopic thermal energy.

The macroscopic energy of the system as a whole is its mechanical energy E_{mech}.

$$E_{sys} = E_{mech} + E_{th}$$

The microscopic motion of the atoms and molecules is kinetic energy K_{micro}. The stretched and compressed bonds have potential energy U_{micro}. Together, these are the system's thermal energy E_{th}.

FIGURE 17.2 The work done by tension can have very different consequences.

(a) Lift at steady speed

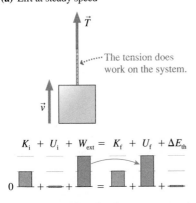

The tension does work on the system.

$$K_i + U_i + W_{ext} = K_f + U_f + \Delta E_{th}$$

The energy transferred to the system goes entirely to the system's mechanical energy.

(b) Drag at steady speed

The tension does work on the system.

$$K_i + U_i + W_{ext} = K_f + U_f + \Delta E_{th}$$

The energy transferred to the system goes entirely to the system's thermal energy.

The point of this example is that the energy transferred to a system can go entirely to the system's mechanical energy, entirely to its thermal energy, or (imagine dragging the object up an incline) some combination of the two. The energy isn't lost, but where it ends up depends on the circumstances.

That Can't Be All

You can transfer energy into a system by the mechanical process of doing work on the system. But that can't be all there is to energy transfer. What happens when you place a pan of water on the stove and light the burner? The water temperature increases, so $\Delta E_{th} > 0$. But no work is done ($W_{ext} = 0$) and there is no change in the water's mechanical energy ($\Delta E_{mech} = 0$). This process clearly violates the energy equation $\Delta E_{mech} + \Delta E_{th} = W_{ext}$. What's wrong?

Nothing is wrong. The energy equation is correct as far as it goes, but it is incomplete. Work is energy transferred in a mechanical interaction, but that is not the only way a system can interact with its environment. Energy can also be transferred between the system and the environment if they have a *thermal interaction*. The energy transferred in a thermal interaction is called *heat*.

The symbol for heat is Q. When heat is included, the energy equation becomes

$$\Delta E_{sys} = \Delta E_{mech} + \Delta E_{th} = W + Q \qquad (17.4)$$

Heat and work are both energy transferred between the system and the environment.

NOTE ▶ We've dropped the subscript "ext" from W. The work that we consider in thermodynamics is *always* the work done by the environment on the system. We won't need to distinguish this work from W_c or W_{diss}, so the subscript is superfluous. ◀

We'll return to Equation 17.4 in Section 17.4 after we look at how work is calculated for ideal-gas processes and at what heat is.

STOP TO THINK 17.1 A gas cylinder and piston are covered with heavy insulation. The piston is pushed into the cylinder, compressing the gas. In this process the gas temperature

a. Increases.
b. Decreases.
c. Doesn't change.
d. There's not sufficient information to tell.

17.2 Work in Ideal-Gas Processes

We introduced the idea of *work* in Chapter 11. **Work** is the energy transferred between a system and the environment when a net force acts on the system over a distance. The process itself is a **mechanical interaction,** meaning that the system and the environment interact via macroscopic pushes and pulls. Loosely speaking, we say that the environment (or a particular force from the environment) "does work" on the system. A system is in **mechanical equilibrium** if there is no net force on the system.

FIGURE 17.3 on the next page reminds you that work can be either positive or negative. **The sign of the work is *not* just an arbitrary convention, nor does it have anything to do with the choice of coordinate system.** The sign of the work tells us which way energy is being transferred.

The pistons in a car engine do work on the air-fuel mixture by compressing it.

FIGURE 17.3 The sign of work.

Work is *positive* when the force is in the direction of motion.
- The force causes the object to speed up.
- Energy is transferred from the environment to the system.
- The system's energy increases.

Work is *negative* when the force is opposite to the motion.
- The force causes the object to slow down.
- Energy is transferred from the system to the environment.
- The system's energy decreases.

In contrast to the mechanical energy or the thermal energy, **work is not a state variable.** That is, work is not a number characterizing the system. Instead, work is the amount of energy that moves between the system and the environment during a mechanical interaction. We can measure the *change* in a state variable, such as a temperature change $\Delta T = T_f - T_i$, but it would make no sense to talk about a "change of work." Consequently, work always appears as W, never as ΔW.

You learned in Chapter 11 how to calculate work. The small amount of work dW done by force $\vec{F}$ as a system moves through the small displacement $d\vec{s}$ is $dW = \vec{F} \cdot d\vec{s}$. If we restrict ourselves to situations where $\vec{F}$ is either parallel or opposite to $d\vec{s}$, then the total work done on the system as it moves from s_i to s_f is

$$W = \int_{s_i}^{s_f} F_s \, ds \qquad (17.5)$$

FIGURE 17.4 The external force does work on the gas as the piston moves.

(a) The gas pushes on the piston with force $\vec{F}_{gas}$.

To keep the piston in place, an external force must be equal and opposite to $\vec{F}_{gas}$.

Pressure p

Piston area A

(b) As the piston moves dx, the external force does work $(F_{ext})_x \, dx$ on the gas.

The volume changes by $dV = A \, dx$ as the piston moves dx.

Let's apply this definition to a gas as it expands or is compressed. FIGURE 17.4a shows a gas cylinder sealed at one end by a movable piston. Force $\vec{F}_{ext}$, perhaps a force supplied by a piston rod, is equal in magnitude and opposite in direction to $\vec{F}_{gas}$. The gas pressure would blow the piston out of the cylinder if the external force weren't there! Using the coordinate system of Figure 17.4a,

$$(F_{ext})_x = -(F_{gas})_x = -pA \qquad (17.6)$$

Suppose the piston moves the small distance dx shown in FIGURE 17.4b. As it does so, the external force (i.e., the environment) does work

$$dW = (F_{ext})_x \, dx = -pA \, dx \qquad (17.7)$$

If dx is positive (the gas expands), then dW is negative. This is because the external force is opposite the displacement. dW is positive if the gas is slightly compressed (negative dx) because the force and the displacement are in the same direction. This is an important idea.

NOTE ▶ The force $\vec{F}_{gas}$ due to the gas pressure inside the cylinder also does work. Because $\vec{F}_{gas} = -\vec{F}_{ext}$, by Newton's third law, the work done by the gas is simply $W_{gas} = -W_{ext}$. To compress the gas, the environment does positive work and the gas does negative work. As the gas expands, W_{gas} is positive and W_{ext} is negative. But the work that appeared in the work-kinetic energy theorem, and now appears in the laws of thermodynamics, is the work done *on* the system by external forces, not the work done *by* the system. It is W_{ext} that tells us whether energy enters the system or leaves the system—by whether it is positive or negative—and that is why we focus our attention on W_{ext} rather than on W_{gas}. ◀

As the piston moves dx, the volume of the gas changes by $dV = A \, dx$. Consequently, Equation 17.7 can be written in terms of the cylinder's volume as

$$dW = -p \, dV \qquad (17.8)$$

If we let the piston move in a slow quasi-static process from initial volume V_i to final volume V_f, the total work done by the environment on the gas is found by integrating Equation 17.8:

$$W = -\int_{V_i}^{V_f} p\, dV \qquad \text{(work done on a gas)} \qquad (17.9)$$

Equation 17.9 is a key result of thermodynamics. Although we used a cylinder to derive Equation 17.9, it turns out to be true for a container of any shape.

NOTE ▶ The pressure of a gas usually changes as the gas expands or contracts. Consequently, p is *not* a constant that can be brought outside the integral. You need to know how the pressure changes with volume before you can carry out the integration. ◀

We can give the work done on a gas a nice geometric interpretation. You learned in Chapter 16 how to represent an ideal-gas process as a curve in the pV diagram. FIGURE 17.5 shows that the work done on a gas is the negative of the area under the pV curve as the volume changes from V_i to V_f. That is

> W = the negative of the area under the pV curve between V_i and V_f

Figure 17.5a shows a process in which a gas *expands* from V_i to a larger volume V_f. The area under the curve is positive, so the environment does a negative amount of work on an expanding gas. Figure 17.5b shows a process in which a gas is compressed to a smaller volume. This one is a little trickier because we have to integrate "backward" along the V-axis. You learned in calculus that integrating from a larger limit to a smaller limit gives a negative result, so the area in Figure 17.5b is a negative area. Consequently, as the minus sign in Equation 17.9 indicates, the environment does positive work on a gas to compress it.

FIGURE 17.5 The work done on a gas is the negative of the area under the curve.

(a) For an *expanding* gas ($V_f > V_i$), the area under the pV curve is positive (integration direction is to the right). Thus the environment does *negative* work on the gas.

(b) For a *compressed* gas ($V_f < V_i$), the area is negative because the integration direction is to the left. Thus the environment does *positive* work on the gas.

EXAMPLE 17.1 The work done on an expanding gas

How much work is done on the gas in the ideal-gas process of FIGURE 17.6?

FIGURE 17.6 The ideal-gas process of Example 17.1.

$(500 \times 10^{-6}\,\text{m}^3)(300{,}000\,\text{Pa}) = 150\,\text{J}$

$\frac{1}{2}(500 \times 10^{-6}\,\text{m}^3)(200{,}000\,\text{Pa}) = 50\,\text{J}$

$(500 \times 10^{-6}\,\text{m}^3)(100{,}000\,\text{Pa}) = 50\,\text{J}$

MODEL The work done on a gas is the negative of the area under the pV curve. The gas is *expanding*, so we expect the work to be negative.

SOLVE As Figure 17.6 shows, the area under the curve can be divided into two rectangles and a triangle. Volumes *must* be converted to SI units of m^3. The total area under the curve is 250 J, so the work done on the gas as it expands is

$$W = -(\text{area under the } pV \text{ curve}) = -250\,\text{J}$$

ASSESS We noted previously that the product Pa m^3 is equivalent to joules. The work is negative, as expected, because the external force pushing on the piston is opposite the direction of the piston's displacement.

Equation 17.9 is the basis for a problem-solving strategy.

> PROBLEM-SOLVING
> STRATEGY 17.1 **Work in ideal-gas processes**
>
> MODEL Assume the gas is ideal and the process is quasi-static.
>
> VISUALIZE Show the process on a pV diagram. Note whether it happens to be one of the basic gas processes: isochoric, isobaric, or isothermal.
>
> SOLVE Calculate the work as the area under the pV curve either geometrically or by carrying out the integration:
>
> $$\text{Work done on the gas } W = -\int_{V_i}^{V_f} p\, dV = -(\text{area under } pV \text{ curve})$$
>
> ASSESS Check your signs.
>
> ■ $W > 0$ when the gas is compressed. Energy is transferred from the environment to the gas.
> ■ $W < 0$ when the gas expands. Energy is transferred from the gas to the environment.
> ■ No work is done if the volume doesn't change. $W = 0$.
>
> Exercise 4

FIGURE 17.7 Calculating the work done during ideal-gas processes.

(a)

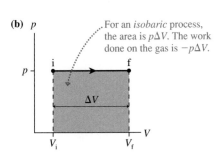

(b)

FIGURE 17.8 An isothermal process.

Isochoric Process

The isochoric process in FIGURE 17.7a is one in which the volume does not change. Consequently,

$$W = 0 \qquad \text{(isochoric process)} \qquad (17.10)$$

An isochoric process is the *only* ideal-gas process in which no work is done.

Isobaric Process

FIGURE 17.7b shows an isobaric process in which the volume changes from V_i to V_f. The rectangular area under the curve is $p\Delta V$, so the work done during this process is

$$W = -p\,\Delta V \qquad \text{(isobaric process)} \qquad (17.11)$$

where $\Delta V = V_f - V_i$. ΔV is positive if the gas expands ($V_f > V_i$), so W is negative. ΔV is *negative* if the gas is compressed ($V_f < V_i$), making W positive.

Isothermal Process

FIGURE 17.8 shows an isothermal process. Here we need to know the pressure as a function of volume before we can integrate Equation 17.9. From the ideal-gas law, $p = nRT/V$. Thus the work on the gas as the volume changes from V_i to V_f is

$$W = -\int_{V_i}^{V_f} p\, dV = -\int_{V_i}^{V_f} \frac{nRT}{V} dV = -nRT \int_{V_i}^{V_f} \frac{dV}{V} \qquad (17.12)$$

where we could take the T outside the integral because temperature is constant during an isothermal process. This is a straightforward integration, giving

$$W = -nRT \int_{V_i}^{V_f} \frac{dV}{V} = -nRT \ln V \Big|_{V_i}^{V_f}$$

$$= -nRT(\ln V_f - \ln V_i) = -nRT \ln\left(\frac{V_f}{V_i}\right) \qquad (17.13)$$

Because $nRT = p_i V_i = p_f V_f$ during an isothermal process, the work is:

$$W = -nRT \ln\left(\frac{V_f}{V_i}\right) = -p_i V_i \ln\left(\frac{V_f}{V_i}\right) = -p_f V_f \ln\left(\frac{V_f}{V_i}\right) \quad (17.14)$$
$$\text{(isothermal process)}$$

Which version of Equation 17.14 is easiest to use will depend on the information you're given. The pressure, volume, and temperature *must* be in SI units.

EXAMPLE 17.2 | **The work of an isothermal compression**

A cylinder contains 7.0 g of nitrogen gas. How much work must be done to compress the gas at a constant temperature of 80°C until the volume is halved?

MODEL This is an isothermal ideal-gas process.

SOLVE Nitrogen gas is N_2, with molar mass $M_{mol} = 28$ g/mol, so 7.0 g is 0.25 mol of gas. The temperature is $T = 353$ K. Although we don't know the actual volume, we do know that $V_f = \frac{1}{2} V_i$. The volume ratio is all we need to calculate the work:

$$W = -nRT \ln\left(\frac{V_f}{V_i}\right)$$
$$= -(0.25 \text{ mol})(8.31 \text{ J/mol K})(353 \text{ K})\ln(1/2) = 508 \text{ J}$$

ASSESS The work is positive because a force from the environment pushes the piston inward to compress the gas.

Work Depends on the Path

FIGURE 17.9a shows two different processes that take a gas from an initial state i to a final state f. Although the initial and final states are the same, the work done during these two processes is *not* the same. **The work done during an ideal-gas process depends on the path followed through the pV diagram.**

You may recall that "work is independent of the path," but that referred to a different situation. In Chapter 11, we found that the work done by a conservative force is independent of the physical path of the object through space. For an ideal-gas process, the "path" is a sequence of thermodynamic states on a pV diagram. It is a figurative path because we can draw a picture of it on a pV diagram, but it is not a literal path.

The path dependence of work has an important implication for multistep processes such as the one shown in FIGURE 17.9b. The total work done on the gas during the process $1 \rightarrow 2 \rightarrow 3$ must be calculated as $W_{1 \text{ to } 3} = W_{1 \text{ to } 2} + W_{2 \text{ to } 3}$. In this case, $W_{1 \text{ to } 2}$ is negative and $W_{2 \text{ to } 3}$ is positive. Trying to compute the work in a single step, using $\Delta V = V_3 - V_1$, would give you the work of a process that goes directly from 1 to 3. The initial and final states are the same, but the work is *not* the same because work depends on the path followed through the pV diagram.

FIGURE 17.9 The work done during an ideal-gas process depends on the path.

(a)

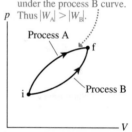

The area under the process A curve is larger than the area under the process B curve. Thus $|W_A| > |W_B|$.

(b)

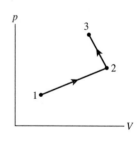

STOP TO THINK 17.2 Two processes take an ideal gas from state 1 to state 3. Compare the work done by process A to the work done by process B.

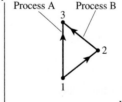

a. $W_A = W_B = 0$
b. $W_A = W_B$ but neither is zero
c. $W_A > W_B$
d. $W_A < W_B$

17.3 Heat

Heat is a more elusive concept than work. We use the word "heat" very loosely in the English language, often as synonymous with *hot*. We might say on a very hot day, "This heat is oppressive." If your apartment is cold, you may say, "Turn up the heat." These expressions date to a time long ago when it was thought that heat was a *substance* with fluid-like properties.

Water

The flame heats the water. The temperature increases.

The spinning paddle does work on the water. The temperature increases.

FIGURE 17.11 The sign of heat.

(a) Positive heat

$Q > 0$

System

$T_{env} > T_{sys}$

(b) Negative heat

$Q < 0$

System

$T_{sys} > T_{env}$

(c) Thermal equilibrium

$Q = 0$

System

$T_{sys} = T_{env}$

Our concept of heat changed with the work of British physicist James Joule in the 1840s. Joule was the first to carry out careful experiments to learn how it is that systems change their temperature. Using experiments like those shown in FIGURE 17.10, Joule found that you can raise the temperature of a beaker of water by two entirely different means:

1. Heating it with a flame, or
2. Doing work on it with a rapidly spinning paddle wheel.

The final state of the water is *exactly* the same in both cases. This implies that heat and work are essentially equivalent. In other words, heat is not a substance. Instead, heat is *energy*. Heat and work, which previously had been regarded as two completely different phenomena, were now seen to be simply two different ways of transferring energy to or from a system.

Thermal Interactions

To be specific, **heat** is the energy transferred between a system and the environment as a consequence of a *temperature difference* between them. Unlike a mechanical interaction in which work is done, heat requires no macroscopic motion of the system. Instead (we'll look at the details in Chapter 18), energy is transferred when the *faster* molecules in the hotter object collide with the *slower* molecules in the cooler object. On average, these collisions cause the faster molecules to lose energy and the slower molecules to gain energy. The net result is that energy is transferred from the hotter object to the colder object. The process itself, whereby energy is transferred between the system and the environment via atomic-level collisions, is called a **thermal interaction.**

When you place a pan of water on the stove, heat is the energy transferred *from* the hotter flame *to* the cooler water. If you place the water in a freezer, heat is the energy transferred from the warmer water to the colder air in the freezer. A system is in **thermal equilibrium** with the environment, or two systems are in thermal equilibrium with each other, if there is no temperature difference.

Like work, **heat is not a state variable. That is, heat is not a property of the system.** Instead, heat is the amount of energy that moves between the system and the environment during a thermal interaction. It would not be meaningful to talk about a "change of heat." Thus heat appears in the energy equation simply as a value Q, never as ΔQ. FIGURE 17.11 shows how to interpret the sign of Q.

NOTE ▶ For both heat and work, a positive value indicates energy being transferred from the environment to the system. Table 17.1 summarizes the similarities and differences between work and heat. ◀

TABLE 17.1 Understanding work and heat

	Work	**Heat**
Interaction:	Mechanical	Thermal
Requires:	Force and displacement	Temperature difference
Process:	Macroscopic pushes and pulls	Microscopic collisions
Positive value:	$W > 0$ when a gas is compressed. Energy is transferred in.	$Q > 0$ when the environment is at a higher temperature than the system. Energy is transferred in.
Negative value:	$W < 0$ when a gas expands. Energy is transferred out.	$Q < 0$ when the system is at a higher temperature than the environment. Energy is transferred out.
Equilibrium:	A system is in mechanical equilibrium when there is no net force or torque on it.	A system is in thermal equilibrium when it is at the same temperature as the environment.

Units of Heat

Heat is energy transferred between the system and the environment. Consequently, the SI unit of heat is the joule. Historically, before the connection between heat and work had been recognized, a unit for measuring heat, the calorie, had been defined as

1 calorie = 1 cal = the quantity of heat needed to change
the temperature of 1 g of water by 1°C

Once Joule established that heat is energy, it was apparent that the calorie is really a unit of energy. In today's SI units, the conversion is

1 cal = 4.186 J

The calorie you know in relation to food is not the same as the heat calorie. The *food calorie,* abbreviated Cal with a capital C, is

1 food calorie = 1 Cal = 1000 cal = 1 kcal = 4186 J

We will not use calories in this textbook, but there are some fields of science and engineering where calories are still widely used. All the calculations you learn to do with joules can equally well be done with calories.

Heat, Temperature, and Thermal Energy

It is important to distinguish among *heat, temperature,* and *thermal energy.* These three ideas are related, but the distinctions among them are crucial. In brief,

- Thermal energy is an energy *of the system* due to the motion of its atoms and molecules. It is a *form* of energy. Thermal energy is a state variable, and it makes sense to talk about how E_{th} changes during a process. The system's thermal energy continues to exist even if the system is isolated and not interacting thermally with its environment.
- Heat is energy transferred *between the system* and the environment as they interact. Heat is *not* a particular form of energy, nor is it a state variable. It makes no sense to talk about how heat changes. $Q = 0$ if a system does not interact thermally with its environment. Heat may cause the system's thermal energy to change, but that doesn't make heat and thermal energy the same.
- Temperature is a state variable that quantifies the "hotness" or "coldness" of a system. We haven't given a precise definition of temperature, but it is related to the thermal energy *per molecule.* A temperature difference is a requirement for a thermal interaction in which heat energy is transferred between the system and the environment.

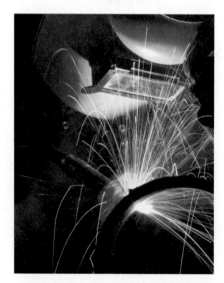

Heat is the energy transferred in a thermal interaction.

It is especially important not to associate an observed temperature increase with heat. Heating a system is one way to change its temperature, but, as Joule showed, not the only way. You can also change the system's temperature by doing work on the system or, as is the case with friction, transforming mechanical energy into thermal energy. **Observing the system tells us nothing about the process by which energy enters or leaves the system.**

STOP TO THINK 17.3   Which one or more of the following processes involves heat?

 a. The brakes in your car get hot when you stop.
 b. A steel block is held over a candle.
 c. You push a rigid cylinder of gas across a frictionless surface.
 d. You push a piston into a cylinder of gas, increasing the temperature of the gas.
 e. You place a cylinder of gas in hot water. The gas expands, causing a piston to rise and lift a weight. The temperature of the gas does not change.

17.4 The First Law of Thermodynamics

Heat was the missing piece that we needed to arrive at a completely general statement of the law of conservation of energy. Restating Equation 17.4, we have

$$\Delta E_{sys} = \Delta E_{mech} + \Delta E_{th} = W + Q$$

Work and heat, two ways of transferring energy between a system and the environment, cause the system's energy to change.

At this point in the text we are not interested in systems that have a macroscopic motion of the system as a whole. Moving macroscopic systems were important to us for many chapters, but now, as we investigate the thermal properties of a system, we would like the system as a whole to rest peacefully on the laboratory bench while we study it. So we will assume, throughout the remainder of Part IV, that $\Delta E_{mech} = 0$.

With this assumption clearly stated, the law of conservation of energy becomes

$$\Delta E_{th} = W + Q \qquad \text{(first law of thermodynamics)} \qquad (17.15)$$

The energy equation, in this form, is called the **first law of thermodynamics** or simply "the first law." The first law is a very general statement about the conservation of energy.

Chapters 10 and 11 introduced the basic energy model. It was called *basic* because it included work but not heat. The first law of thermodynamics has included heat, but it excludes situations where the mechanical energy changes. FIGURE 17.12 is a pictorial representation of the **thermodynamic energy model** described by the first law. Work and heat are energies transferred between the system and the environment. Energy added to the system (W or Q positive) increases the system's thermal energy ($\Delta E_{th} > 0$). Likewise, the thermal energy decreases when energy is removed from the system.

Two comments are worthwhile:

1. The first law of thermodynamics doesn't tell us anything about the value of E_{th}, only how E_{th} changes. Doing 1 J of work changes the thermal energy by $\Delta E_{th} = 1$ J regardless of whether $E_{th} = 10$ J or 10,000 J.
2. The system's thermal energy isn't the only thing that changes. Work or heat that changes the thermal energy also changes the pressure, volume, temperature, and other state variables. The first law tells us only about ΔE_{th}. Other laws and relationships must be used to learn how the other state variables change.

Three Special Ideal-Gas Processes

There are three ideal-gas processes in which one of the terms in the first law—ΔE_{th}, W, or Q—is zero. To investigate these processes, FIGURE 17.13 shows a gas cylinder with three special properties:

- You can keep the gas volume from changing by inserting the locking pin into the piston. Without the pin, the piston can slide up or down. The piston is massless, frictionless, and insulated.
- You can change the gas pressure by adding or removing masses on top of the piston. Work is done as the piston moves the masses up and down.
- You can warm or cool the gas by placing the cylinder above a flame or on a block of ice. The thin bottom of the cylinder is the only surface through which heat energy can be transferred.

You learned in Chapter 16 (see Figure 16.9) that the gas pressure when the piston "floats" is determined by the atmospheric pressure and by the total mass M on the piston:

$$p_{gas} = p_{atmos} + \frac{Mg}{A} \qquad (17.16)$$

The pressure doesn't change as the piston moves unless you change the mass. This is a particularly important point to understand. Equation 17.16 is *not* valid when the piston is locked. The pressure with the piston locked could be either higher or lower than the value found with Equation 17.16.

FIGURE 17.12 The thermodynamic energy model.

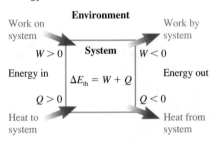

FIGURE 17.13 The gas can be heated and have work done on it.

Masses determine the gas pressure. Work is done as the masses move up and down.

Insulation

Piston

A locking pin fixes the gas volume.

Gas

Thin bottom

Heat energy can be transferred into or out of the gas.

Flame

Ice

An isochoric cooling process ($W = 0$): No work is done in an isochoric (constant volume) process because the piston doesn't move. To cool the gas without doing work:

■ Insert the locking pin so that the volume cannot change.
■ Place the cylinder on the block of ice. Heat energy will be transferred from the gas to the ice, causing the gas temperature and pressure to fall.
■ Remove the cylinder from the ice when the desired pressure is reached.
■ Remove masses from the piston until the total mass M balances the new gas pressure. This step must be done before removing the locking pin; otherwise, the piston will move when the pin is removed.
■ Remove the locking pin.

Figure 17.7a showed the pV diagram. The final point is on a lower isotherm than the initial point, so $T_f < T_i$. No work was done, but heat energy was transferred out of the gas ($Q < 0$) and the thermal energy of the gas decreased ($\Delta E_{th} < 0$) as the temperature fell. FIGURE 17.14 shows this result on a first-law bar chart.

An isothermal expansion ($\Delta E_{th} = 0$): The thermal energy does not change in an isothermal process because the temperature of the gas doesn't change. To expand the gas without changing its thermal energy:

■ Place the cylinder over the flame. Heat energy will be transferred to the gas, and the gas will begin to expand.
■ The product pV must remain constant during an isothermal process. Slowly remove masses from the piston to reduce the pressure as the volume increases. The temperature remains constant as heat energy from the flame balances the negative work done on the gas as it expands.
■ Remove the cylinder from the flame when the gas reaches the desired volume.

Figure 17.8 showed the pV diagram, and FIGURE 17.15 is the first-law bar chart. The temperature doesn't change in an isothermal process ($\Delta T = 0$), hence the thermal energy cannot change ($\Delta E_{th} = 0$). Heat energy is transferred to the gas, but that energy is used to do work (the piston lifts the masses) rather than to increase the temperature. Here "do work" means that the gas is doing work ($W_{gas} > 0$), so the external work *on* the gas—the W in the first law—is negative.

NOTE ▶ It is surprising, but true, that we can heat the system without changing its temperature. But to do so, we must have a process in which the energy coming into the system as heat is exactly balanced by the energy leaving the system as work. **The important point is that $\Delta T = 0$ does *not* mean $Q = 0$.** ◀

An adiabatic compression ($Q = 0$): A process in which no heat energy is transferred between the system and the environment is called an **adiabatic process.** Although the system cannot have thermal interactions with its environment, it can still have mechanical interactions as the insulated piston pushes or pulls on the gas. To compress the gas without heat:

■ Add insulation beneath the cylinder.
■ Slowly add masses to the piston, increasing the pressure. The piston will slowly descend, compressing the gas and decreasing its volume.
■ Stop adding masses when the gas reaches the desired volume.

$Q = 0$ in an adiabatic process, so the first law $\Delta E_{th} = W + Q$ can be satisfied only if $\Delta E_{th} = W$. This information is shown on the first-law bar chart of FIGURE 17.16.

NOTE ▶ Just because the system is well insulated—thermally isolated from the environment—does not mean its temperature remains constant. An adiabatic compression uses work to increase the temperature of the gas. Similarly, an adiabatic expansion lowers the temperature of the gas. **The important point is that $Q = 0$ does *not* mean $\Delta T = 0$.** ◀

FIGURE 17.14 A first-law bar chart for a process that does no work.

Thermal energy has decreased by the amount of energy that left the system as heat.

FIGURE 17.15 A first-law bar chart for a process that doesn't change the thermal energy.

The energy that enters the system as heat leaves as work. The thermal energy is unchanged.

FIGURE 17.16 A first-law bar chart for a process that transfers no heat energy.

Energy that enters the system as work increases the thermal energy—and thus the temperature.

We'll examine adiabatic gas processes and their pV curve later in the chapter. For now, make sure you understand which quantities are zero and which aren't in these three special processes.

STOP TO THINK 17.4 Which first-law bar chart describes the process shown in the pV diagram?

(a)   (b)   (c)   (d)

17.5 Thermal Properties of Matter

Heat and work are equivalent in the sense that the change of the system is *exactly the same* whether you transfer heat energy to it or do an equal amount of work on it. Adding energy to the system, or removing it, changes the system's thermal energy.

What happens to a system when you change its thermal energy? In this section we'll consider two distinct possibilities:

- The temperature of the system changes.
- The system undergoes a phase change, such as melting or freezing.

Temperature Change and Specific Heat

TABLE 17.2 Specific heats and molar specific heats of solids and liquids

Substance	c (J/kg K)	C (J/mol K)
Solids		
Aluminum	900	24.3
Copper	385	24.4
Iron	449	25.1
Gold	129	25.4
Lead	128	26.5
Ice	2090	37.6
Liquids		
Ethyl alcohol	2400	110.4
Mercury	140	28.1
Water	4190	75.4

Suppose you do an experiment in which you add energy to water, either by doing work on it or by transferring heat to it. Either way, you will find that adding 4190 J of energy raises the temperature of 1 kg of water by 1 K. If you were fortunate enough to have 1 kg of gold, you would need to add only 129 J of energy to raise its temperature by 1 K.

The amount of energy that raises the temperature of 1 kg of a substance by 1 K is called the **specific heat** of that substance. The symbol for specific heat is c. Water has specific heat $c_{water} = 4190$ J/kg K. The specific heat of gold is $c_{gold} = 129$ J/kg K. Specific heat depends only on the material from which an object is made. Table 17.2 provides some specific heats for common liquids and solids.

NOTE ▶ The term *specific heat* does not use the word "heat" in the way that we have defined it. Specific heat is an old idea, dating back to the days of the caloric theory when heat was thought to be a substance contained in the object. The term has continued in use even though our understanding of heat has changed. ◀

If energy c is required to raise the temperature of 1 kg of a substance by 1 K, then energy Mc is needed to raise the temperature of mass M by 1 K and $(Mc)\Delta T$ is needed to raise the temperature of mass M by ΔT. In other words, the thermal energy of the system changes by

$$\Delta E_{th} = Mc\Delta T \quad \text{(temperature change)} \quad (17.17)$$

when its temperature changes by ΔT. ΔE_{th} can be either positive (thermal energy increases as the temperature goes up) or negative (thermal energy decreases as the temperature goes down). Recall that uppercase M is used for the mass of an entire system while lowercase m is reserved for the mass of an atom or molecule.

NOTE ▶ In practice, ΔT is usually measured in °C. But the Kelvin and the Celsius temperature scales have the same step size, so ΔT in K has exactly the same numerical value as ΔT in °C. Thus

- You do not need to convert temperatures from °C to K if you need only a temperature *change* ΔT.
- You do need to convert anytime you need the actual temperature T. ◀

The first law of thermodynamics, $\Delta E_{th} = W + Q$, allows us to write Equation 17.17 as $Mc\Delta T = W + Q$. In other words, **we can change the system's temperature either by heating it or by doing an equivalent amount of work on it.** In working with solids and liquids, we almost always change the temperature by heating. If $W = 0$, which we will assume for the rest of this section, then the heat energy needed to bring about a temperature change ΔT is

$$Q = Mc\Delta T \qquad \text{(temperature change)} \qquad (17.18)$$

Because $\Delta T = \Delta E_{th}/Mc$, it takes more energy to change the temperature of a substance with a large specific heat than to change the temperature of a substance with a small specific heat. You can think of specific heat as measuring the *thermal inertia* of a substance. Metals, with small specific heats, warm up and cool down quickly. A piece of aluminum foil can be safely held within seconds of removing it from a hot oven. Water, with a very large specific heat, is slow to warm up and slow to cool down. This is fortunate for us. The large thermal inertia of water is essential for the biological processes of life. We wouldn't be here studying physics if water had a small specific heat!

EXAMPLE 17.3   **Running a fever**

A 70 kg student catches the flu, and his body temperature increases from 37.0°C (98.6°F) to 39.0°C (102.2°F). How much energy is required to raise his body's temperature? The specific heat of a mammalian body is 3400 J/kg K, nearly that of water because mammals are mostly water.

MODEL Energy is supplied by the chemical reactions of the body's metabolism. These exothermic reactions transfer heat to the body. Normal metabolism provides enough heat energy to offset energy losses (radiation, evaporation, etc.) while maintaining

a normal body temperature of 37°C. We need to calculate the additional energy needed to raise the body's temperature by 2.0°C, or 2.0 K.

SOLVE The necessary heat energy is

$$Q = Mc\Delta T = (70 \text{ kg})(3400 \text{ J/kg K})(2.0 \text{ K}) = 4.8 \times 10^5 \text{ J}$$

ASSESS This appears to be a lot of energy, but a joule is actually a very small amount of energy. It is only 110 Cal, approximately the energy gained by eating an apple.

The **molar specific heat** is the amount of energy that raises the temperature of 1 mol of a substance by 1 K. We'll use an uppercase C for the molar specific heat. The heat energy needed to bring about a temperature change ΔT of n moles of substance is

$$Q = nC\Delta T \qquad (17.19)$$

Molar specific heats are listed in Table 17.2. Look at the five elemental solids (excluding ice). All have C very near 25 J/mol K. If we were to expand the table, we would find that most elemental solids have $C \approx 25$ J/mol K. This can't be a coincidence, but what is it telling us? This is a puzzle we will address in Chapter 18, where we will explore thermal energy at the atomic level.

FIGURE 17.17 The temperature of a
system that is heated at a steady rate.

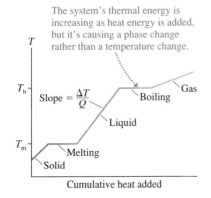

The system's thermal energy is
increasing as heat energy is added,
but it's causing a phase change
rather than a temperature change.

Cumulative heat added

Lava—molten rock—undergoes a phase
change when it contacts the much colder
water. This is one way in which new
islands are formed.

Phase Change and Heat of Transformation

Suppose you start with a system in its solid phase and heat it at a steady rate. FIGURE 17.17, which you saw in Chapter 16, shows how the system's temperature changes. At first, the temperature increases linearly. This is not hard to understand because Equation 17.18 can be written

$$\text{slope of the } T\text{-versus-}Q \text{ graph} = \frac{\Delta T}{Q} = \frac{1}{Mc} \qquad (17.20)$$

The slope of the graph depends inversely on the system's specific heat. A constant specific heat implies a constant slope and thus a linear graph. In fact, you can measure c from such a graph.

NOTE ▶ The different slopes indicate that the solid, liquid, and gas phases of a substance have different specific heats. ◀

But there are times, shown as horizontal line segments, during which heat is being transferred to the system but the temperature isn't changing. These are *phase changes*. The thermal energy continues to increase during a phase change, but the additional energy goes into breaking molecular bonds rather than speeding up the molecules. **A phase change is characterized by a change in thermal energy without a change in temperature.**

The amount of heat energy that causes 1 kg of a substance to undergo a phase change is called the **heat of transformation** of that substance. For example, laboratory experiments show that 333,000 J of heat are needed to melt 1 kg of ice at 0°C. The symbol for heat of transformation is L. The heat required for the entire system of mass M to undergo a phase change is

$$Q = ML \qquad \text{(phase change)} \qquad (17.21)$$

Heat of transformation is a generic term that refers to any phase change. Two specific heats of transformation are the **heat of fusion** L_f, the heat of transformation between a solid and a liquid, and the **heat of vaporization** L_v, the heat of transformation between a liquid and a gas. The heat needed for these phase changes is

$$Q = \begin{cases} \pm ML_f & \text{melt/freeze} \\ \pm ML_v & \text{boil/condense} \end{cases} \qquad (17.22)$$

where the $\pm$ indicates that heat must be *added* to the system during melting or boiling but *removed* from the system during freezing or condensing. **You must explicitly include the minus sign when it is needed.**

Table 17.3 gives the heats of transformation of a few substances. Notice that the heat of vaporization is always much larger than the heat of fusion. We can understand this. Melting breaks just enough molecular bonds to allow the system to lose rigidity and flow. Even so, the molecules in a liquid remain close together and loosely bonded. Vaporization breaks all bonds completely and sends the molecules flying apart. This process requires a larger increase in the thermal energy and thus a larger quantity of heat.

TABLE 17.3 Melting/boiling temperatures and heats of transformation

Substance	T_m (°C)	L_f (J/kg)	T_b (°C)	L_v (J/kg)
Nitrogen (N$_2$)	−210	0.26×10^5	−196	1.99×10^5
Ethyl alcohol	−114	1.09×10^5	78	8.79×10^5
Mercury	−39	0.11×10^5	357	2.96×10^5
Water	0	3.33×10^5	100	22.6×10^5
Lead	328	0.25×10^5	1750	8.58×10^5

EXAMPLE 17.4 **Melting wax**

An insulated jar containing 200 g of solid candle wax is placed on a hot plate that supplies heat energy to the wax at the rate of 220 J/s. The wax temperature is measured every 30 s, yielding the following data:

Time (s)	Temperature (°C)	Time (s)	Temperature (°C)
0	20.0	180	70.5
30	31.7	210	70.5
60	42.2	240	70.6
90	55.0	270	70.5
120	64.7	300	70.4
150	70.4	330	74.5

What are the specific heat of the solid wax, the melting point, and the wax's heat of fusion?

MODEL The wax is in an insulated jar, so assume that heat loss to the environment is negligible.

VISUALIZE Heat energy is being supplied at the rate of 220 J/s, so the total heat energy that has been transferred into the wax at time t is $Q = 220t$ J. FIGURE 17.18 shows the temperature graphed against the cumulative heat Q, although notice that the horizontal axis is in kJ, not J. The initial linear slope corresponds to raising the wax's temperature to the melting point. Temperature remains constant during a phase change, even though the sample is still being heated, so the horizontal section of the graph is when the wax is melting. The temperature increase at the end shows that the temperature of the liquid wax is beginning to rise after melting is complete.

SOLVE From $Q = Mc\Delta T$, the slope of the T-versus-Q graph is $\Delta T/Q = 1/Mc$. The experimental slope of the best-fit line is 1.708°C/kJ = 0.001708 K/J. Thus the specific heat of the solid wax is

$$c = \frac{1}{M \times \text{slope}} = \frac{1}{(0.200 \text{ kg})(0.001708 \text{ K/J})} = 2930 \text{ J/kg K}$$

From the table, we see that the melting temperature—which remains constant during the phase change—is 70.5°C. The heat required for the phase change is $Q = ML_f$, so the heat of fusion is $L_f = Q/M$. With data recorded only every 30 s, it's not exactly clear when the melting began and when it ended. The extension of the initial slope shows that the temperature reached the melting point about halfway between 120 s and 150 s, so the melting started at about 135 s. We'll assume it was complete about halfway between 300 s and 330 s, or at about 315 s. Thus the melting took 180 s, during which, at 220 J/s, 39,600 J of heat energy was transferred from the hot plate to the wax. With this value of Q, the heat of fusion is

$$L_f = \frac{Q}{M} = \frac{39,600 \text{ J}}{0.200 \text{ kg}} = 2.0 \times 10^5 \text{ J/kg}$$

ASSESS Both the specific heat and the heat of fusion are similar to values in Tables 17.2 and 17.3, which gives us confidence in our results.

FIGURE 17.18 The heating curve of the wax.

STOP TO THINK 17.5 Objects A and B are brought into close thermal contact with each other, but they are well isolated from their surroundings. Initially $T_A = 0°C$ and $T_B = 100°C$. The specific heat of A is less than the specific heat of B. The two objects will soon reach a common final temperature T_f. The final temperature is

a. $T_f > 50°C$ b. $T_f = 50°C$ c. $T_f < 50°C$

17.6 Calorimetry

At one time or another you've probably put an ice cube into a hot drink to cool it quickly. You were engaged, in a somewhat trial-and-error way, in a practical aspect of heat transfer known as **calorimetry.**

FIGURE 17.19 on the next page shows two systems thermally interacting with each other but isolated from everything else. Suppose they start at different temperatures T_1 and T_2. As you know from experience, heat energy will be transferred from the hotter

FIGURE 17.19 Two systems interact thermally.

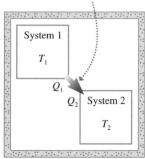

Heat energy is transferred from system 1 to system 2. Energy conservation requires

$$|Q_1| = |Q_2|$$

Opposite signs mean that

$$Q_{net} = Q_1 + Q_2 = 0$$

to the colder system until they reach a common final temperature T_f. The systems will then be in thermal equilibrium and the temperature will not change further.

The insulation prevents any heat energy from being transferred to or from the environment, so energy conservation tells us that any energy leaving the hotter system must enter the colder system. That is, the systems *exchange* energy with no net loss or gain. The concept is straightforward, but to state the idea mathematically we need to be careful with signs.

Let Q_1 be the energy transferred to system 1 as heat. Similarly, Q_2 is the energy transferred to system 2. The fact that the systems are merely exchanging energy can be written $|Q_1| = |Q_2|$. That is, the energy *lost* by the hotter system is the energy *gained* by the colder system. Thus Q_1 and Q_2 have opposite signs: $Q_1 = -Q_2$. No energy is exchanged with the environment, hence it makes more sense to write this relationship as

$$Q_{net} = Q_1 + Q_2 = 0 \tag{17.23}$$

This idea is not limited to the interaction of only two systems. If three or more systems are combined in isolation from the rest of their environment, each at a different initial temperature, they will all come to a common final temperature that can be found from the relationship

$$Q_{net} = Q_1 + Q_2 + Q_3 + \cdots = 0 \tag{17.24}$$

NOTE ▶ The signs are very important in calorimetry problems. ΔT is always $T_f - T_i$, so ΔT and Q are negative for any system whose temperature decreases. The proper sign of Q for any phase change must be supplied *by you*, depending on the direction of the phase change. ◀

PROBLEM-SOLVING
STRATEGY 17.2 **Calorimetry problems**

MODEL Identify the interacting systems. Assume that they are isolated from the larger environment.

VISUALIZE List known information and identify what you need to find. Convert all quantities to SI units.

SOLVE The mathematical representation, which is a statement of energy conservation, is

$$Q_{net} = Q_1 + Q_2 + \cdots = 0$$

- For systems that undergo a temperature change, $Q = Mc(T_f - T_i)$. Be sure to have the temperatures T_i and T_f in the correct order.
- For systems that undergo a phase change, $Q = \pm ML$. Supply the correct sign by observing whether energy enters or leaves the system.
- Some systems may undergo a temperature change *and* a phase change. Treat the changes separately. The heat energy is $Q = Q_{\Delta T} + Q_{phase}$.

ASSESS Is the final temperature in the middle? T_f that is higher or lower than all initial temperatures is an indication that something is wrong, usually a sign error.

Exercise 15

NOTE ▶ You may have learned to solve calorimetry problems in other courses by writing $Q_{gained} = Q_{lost}$. That is, by balancing heat gained with heat lost. That approach works in simple problems, but it has two drawbacks. First, you often have to "fudge" the signs to make them work. Second, and more serious, you can't extend this approach to a problem with three or more interacting systems. Using $Q_{net} = 0$ is much preferred. ◀

EXAMPLE 17.5 **Calorimetry with a phase change**

Your 500 mL soda is at 20°C, room temperature, so you add 100 g of ice from the −20°C freezer. Does all the ice melt? If so, what is the final temperature? If not, what fraction of the ice melts? Assume that you have a well-insulated cup.

MODEL We have a thermal interaction between the soda, which is essentially water, and the ice. We need to distinguish between three possible outcomes. If all the ice melts, then $T_f > 0°C$. It's also possible that the soda will cool to 0°C before all the ice has melted, leaving the ice and liquid in equilibrium at 0°C. A third possibility is that the soda will freeze solid before the ice warms up to 0°C. That seems unlikely here, but there are situations, such as the pouring of molten metal out of furnaces, when all the liquid does solidify. We need to distinguish between these before knowing how to proceed.

VISUALIZE All the initial temperatures, masses, and specific heats are known. The final temperature of the combined soda + ice system is unknown.

SOLVE Let's first calculate the heat needed to melt all the ice and leave it as liquid water at 0°C. To do so, we must warm the ice to 0°C, then change it to water. The heat input for this two-stage process is

$$Q_{melt} = M_i c_i (20 \text{ K}) + M_i L_f = 37,500 \text{ J}$$

where L_f is the heat of fusion of water. It is used as a *positive* quantity because we must *add* heat to melt the ice. Next, let's calculate how much heat energy will leave the soda if it cools all the way to 0°C. The volume is $V = 500 \text{ mL} = 5.00 \times 10^{-4} \text{ m}^3$ and thus the mass is $M_s = \rho V = 0.500 \text{ kg}$. The heat is

$$Q_{cool} = M_s c_w (-20 \text{ K}) = -41,900 \text{ J}$$

where $\Delta T = -20 \text{ K}$ because the temperature decreases. Because $|Q_{cool}| > Q_{melt}$, the soda has sufficient energy to melt all the ice. Hence the final state will be all liquid at $T_f > 0$. (Had we found $|Q_{cool}| < Q_{melt}$, then the final state would have been an ice-liquid mixture at 0°C.)

Energy conservation requires $Q_{ice} + Q_{soda} = 0$. The heat Q_{ice} consists of three terms: warming the ice to 0°C, melting the ice to water at 0°C, then warming the 0°C water to T_f. The mass will still be M_i in the last of these steps because it is the "ice system," but we need to use the specific heat of *liquid water*. Thus

$$Q_{ice} + Q_{soda} = [M_i c_i (20 \text{ K}) + M_i L_f + M_i c_w (T_f - 0°C)]$$
$$+ M_s c_w (T_f - 20°C) = 0$$

We've already done part of the calculation, allowing us to write

$$37,500 \text{ J} + M_i c_w (T_f - 0°C) + M_s c_w (T_f - 20°C) = 0$$

Solving for T_f gives

$$T_f = \frac{20 M_s c_w - 37,500}{M_i c_w + M_s c_w} = 1.7°C$$

ASSESS As expected, the soda has been cooled to nearly the freezing point.

EXAMPLE 17.6 **Three interacting systems**

A 200 g piece of iron at 120°C and a 150 g piece of copper at −50°C are dropped into an insulated beaker containing 300 g of ethyl alcohol at 20°C. What is the final temperature?

MODEL Here you can't use a simple $Q_{gained} = Q_{lost}$ approach because you don't know whether the alcohol is going to warm up or cool down.

VISUALIZE All the initial temperatures, masses, and specific heats are known. We need to find the final temperature.

SOLVE Energy conservation requires

$$Q_i + Q_c + Q_e = M_i c_i (T_f - 120°C) + M_c c_c (T_f - (-50°C))$$
$$+ M_e c_e (T_f - 20°C) = 0$$

Solving for T_f gives

$$T_f = \frac{120 M_i c_i - 50 M_c c_c + 20 M_e c_e}{M_i c_i + M_c c_c + M_e c_e} = 25.7°C$$

ASSESS The temperature is between the initial iron and copper temperatures, as expected. It turns out that the alcohol warms up ($Q_e > 0$), but we had no way to know this without doing the calculation.

17.7 The Specific Heats of Gases

Specific heats are given in Table 17.2 for solids and liquids. Gases are harder to characterize because the heat required to cause a specified temperature change depends on the *process* by which the gas changes state.

FIGURE 17.20 shows two isotherms on the pV diagram for a gas. Processes A and B, which start on the T_i isotherm and end on the T_f isotherm, have the *same* temperature change $\Delta T = T_f - T_i$. But process A, which takes place at constant volume, requires a *different* amount of heat than does process B, which occurs at constant pressure. The reason is that work is done in process B but not in process A. This is a situation that we are now equipped to analyze.

It is useful to define two different versions of the specific heat of gases, one for constant-volume (isochoric) processes and one for constant-pressure (isobaric) processes. We will define these as molar specific heats because we usually do gas

FIGURE 17.20 Processes A and B have the same ΔT and the same ΔE_{th}, but they require different amounts of heat.

TABLE 17.4 Molar specific heats of gases (J/mol K)

Gas	C_P	C_V	$C_P - C_V$
Monatomic Gases			
He	20.8	12.5	8.3
Ne	20.8	12.5	8.3
Ar	20.8	12.5	8.3
Diatomic Gases			
H_2	28.7	20.4	8.3
N_2	29.1	20.8	8.3
O_2	29.2	20.9	8.3

calculations using moles instead of mass. The quantity of heat needed to change the temperature of n moles of gas by ΔT is

$$Q = nC_V\Delta T \quad \text{(temperature change at constant volume)}$$
$$Q = nC_P\Delta T \quad \text{(temperature change at constant pressure)} \tag{17.25}$$

where C_V is the **molar specific heat at constant volume** and C_P is the **molar specific heat at constant pressure**. Table 17.4 gives the values of C_V and C_P for a few common monatomic and diatomic gases. The units are J/mol K.

NOTE ▶ Equations 17.25 apply to two specific ideal-gas processes. In a general gas process, for which neither p nor V is constant, we have no direct way to relate Q to ΔT. In that case, the heat must be found indirectly from the first law as $Q = \Delta E_{th} - W$. ◀

EXAMPLE 17.7 **Heating and cooling a gas**

Three moles of O_2 gas are at 20.0°C. 600 J of heat energy are transferred to the gas at constant pressure, then 600 J are removed at constant volume. What is the final temperature? Show the process on a pV diagram.

MODEL O_2 is a diatomic ideal gas. The gas is heated as an isobaric process, then cooled as an isochoric process.

SOLVE The heat transferred during the constant-pressure process causes a temperature rise

$$\Delta T = T_2 - T_1 = \frac{Q}{nC_P} = \frac{600\,\text{J}}{(3.0\,\text{mol})(29.2\,\text{J/mol K})} = 6.8°\text{C}$$

where C_P for oxygen was taken from Table 17.4. Heating leaves the gas at temperature $T_2 = T_1 + \Delta T = 26.8°\text{C}$. The temperature then falls as heat is removed during the constant-volume process:

$$\Delta T = T_3 - T_2 = \frac{Q}{nC_V} = \frac{-600\,\text{J}}{(3.0\,\text{mol})(20.9\,\text{J/mol K})} = -9.5°\text{C}$$

We used a *negative* value for Q because heat energy is transferred from the gas to the environment. The final temperature of the gas is $T_3 = T_2 + \Delta T = 17.3°\text{C}$. **FIGURE 17.21** shows the process on a

pV diagram. The gas expands (moves horizontally on the diagram) as heat is added, then cools at constant volume (moves vertically on the diagram) as heat is removed.

FIGURE 17.21 The pV diagram for Example 17.7.

ASSESS The final temperature is lower than the initial temperature because $C_P > C_V$.

EXAMPLE 17.8 **Calorimetry with a gas and a solid**

The interior volume of a 200 g hollow aluminum box is 800 cm³. The box contains nitrogen gas at STP. A 20 cm³ block of copper at a temperature of 300°C is placed inside the box, then the box is sealed. What is the final temperature?

MODEL This example has three interacting systems: the aluminum box, the nitrogen gas, and the copper block. They must all come to a common final temperature T_f.

VISUALIZE The box and gas have the same initial temperature: $T_{Al} = T_{N2} = 0°\text{C}$. The box doesn't change size, so this is a constant-volume process. The final temperature is unknown.

SOLVE Although one of the systems is now a gas, the calorimetry equation $Q_{net} = Q_{Al} + Q_{N2} + Q_{Cu} = 0$ is still appropriate. In this case,

$$Q_{net} = m_{Al}c_{Al}(T_f - T_{Al}) + n_{N2}C_V(T_f - T_{N2})$$
$$+ m_{Cu}c_{Cu}(T_f - T_{Cu}) = 0$$

Notice that we used masses and specific heats for the solids but moles and the molar specific heat for the gas. We used C_V because this is a constant-volume process. Solving for T_f gives

$$T_f = \frac{m_{Al}c_{Al}T_{Al} + n_{N2}C_VT_{N2} + m_{Cu}c_{Cu}T_{Cu}}{m_{Al}c_{Al} + n_{N2}C_V + m_{Cu}c_{Cu}}$$

The specific heat values are found in Tables 17.2 and 17.4. The mass of the copper is

$$m_{Cu} = \rho_{Cu}V_{Cu} = (8920\,\text{kg/cm}^3)(20 \times 10^{-6}\,\text{m}^3) = 0.178\,\text{kg}$$

The number of moles of the gas is found from the ideal-gas law, using the initial conditions. Notice that inserting the copper block *displaces* 20 cm³ of gas; hence the gas volume is only $V = 780\,\text{cm}^3 = 7.80 \times 10^{-4}\,\text{m}^3$. Thus

$$n_{N2} = \frac{pV}{RT} = 0.0348\,\text{mol}$$

Computing the final temperature gives $T_f = 83°\text{C}$.

C_P and C_V

You may have noticed two curious features in Table 17.4. First, the molar specific heats of monatomic gases are *all alike*. And the molar specific heats of diatomic gases, while different from monatomic gases, are again *very nearly alike*. We saw a similar feature in Table 17.2 for the molar specific heats of solids. Second, the *difference* $C_P - C_V = 8.3$ J/mol K is the same in every case. And, most puzzling of all, the value of $C_P - C_V$ appears to be equal to the universal gas constant R! Why should this be?

The relationship between C_V and C_P hinges on one crucial idea: ΔE_{th}, **the change in the thermal energy of a gas, is the same for *any* two processes that have the same ΔT.** The thermal energy of a gas is associated with temperature, so any process that changes the gas temperature from T_i to T_f has the same ΔE_{th} as any other process that goes from T_i to T_f. Furthermore, the first law $\Delta E_{th} = Q + W$ tells us that a gas cannot distinguish between heat and work. The system's thermal energy changes in response to energy added to or removed from the system, but the response of the gas is the same whether you heat the system, do work on the system, or do some combination of both. Thus **any two processes that change the thermal energy of the gas by ΔE_{th} will cause the same temperature change ΔT.**

With that in mind, look back at Figure 17.20. Both gas processes have the same ΔT, so both have the same value of ΔE_{th}. Process A is an isochoric process in which no work is done (the piston doesn't move), so the first law for this process is

$$(\Delta E_{th})_A = W + Q = 0 + Q_{const\ vol} = nC_V\Delta T \qquad (17.26)$$

Process B is an isobaric process. You learned earlier that the work done on the gas during an isobaric process is $W = -p\Delta V$. Thus

$$(\Delta E_{th})_B = W + Q = -p\Delta V + Q_{const\ press} = -p\Delta V + nC_P\Delta T \quad (17.27)$$

$(\Delta E_{th})_B = (\Delta E_{th})_A$ because both have the same ΔT, so we can equate the right sides of Equations 17.26 and 17.27:

$$-p\Delta V + nC_P\Delta T = nC_V\Delta T \qquad (17.28)$$

For the final step, we can use the ideal-gas law $pV = nRT$ to relate ΔV and ΔT during process B. For any gas process,

$$\Delta(pV) = \Delta(nRT) \qquad (17.29)$$

For a constant-pressure process, where p is constant, Equation 17.29 becomes

$$p\Delta V = nR\Delta T \qquad (17.30)$$

Substituting this expression for $p\Delta V$ into Equation 17.28 gives

$$-nR\Delta T + nC_P\Delta T = nC_V\Delta T \qquad (17.31)$$

The $n\Delta T$ cancels, and we are left with

$$C_P = C_V + R \qquad (17.32)$$

This result, which applies to all ideal gases, is exactly what we see in the data of Table 17.4.

But that's not the only conclusion we can draw. Equation 17.26 found that $\Delta E_{th} = nC_V\Delta T$ for a constant-volume process. But we had just noted that ΔE_{th} is the same for *all* gas processes that have the same ΔT. Consequently, this expression for ΔE_{th} is equally true for any other process. That is

$$\Delta E_{th} = nC_V\Delta T \qquad \text{(any ideal-gas process)} \qquad (17.33)$$

Compare this result to Equations 17.25. We first made a distinction between constant-volume and constant-pressure processes, but now we're saying that Equation 17.33 is true for any process. Are we contradicting ourselves? No, the difference lies in what you need to calculate.

- The change in thermal energy when the temperature changes by ΔT is the same for any process. That's Equation 17.33.
- The *heat* required to bring about the temperature change depends on what the process is. That's Equations 17.25. An isobaric process requires more heat than an isochoric process that produces the same ΔT.

The reason for the difference is seen by writing the first law as $Q = \Delta E_{th} - W$. In an isochoric process, where $W = 0$, *all* the heat input is used to increase the gas temperature. But in an isobaric process, some of the energy that enters the system as heat then leaves the system as work ($W < 0$) done by the expanding gas. Thus more heat is needed to produce the same ΔT.

Heat Depends on the Path

FIGURE 17.22 Is the heat input along these two paths the same or different?

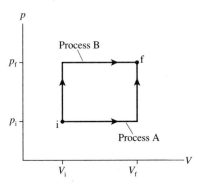

Consider the two ideal-gas processes shown in FIGURE 17.22. Even though the initial and final states are the same, the heat added during these two processes is *not* the same. We can use the first law $\Delta E_{th} = W + Q$ to see why.

The thermal energy is a state variable. That is, its value depends on the state of the gas, not the process by which the gas arrived at that state. Thus $\Delta E_{th} = E_{th\,f} - E_{th\,i}$ is the same for both processes. If ΔE_{th} is the same for processes A and B, then $W_A + Q_A = W_B + Q_B$.

You learned in Section 17.2 that the work done during an ideal-gas process depends on the path in the pV diagram. There's more area under the process B curve, so $|W_B| > |W_A|$. Both values of W are negative because the gas expands, so W_B is more negative than W_A. Consequently, $W_A + Q_A$ can equal $W_B + Q_B$ only if $Q_B > Q_A$. **The heat added or removed during an ideal-gas process depends on the path followed through the pV diagram.**

Adiabatic Processes

FIGURE 17.23 The relationship of three important processes to the first law of thermodynamics.

Section 17.4 introduced the idea of an *adiabatic process,* a process in which no heat energy is transferred ($Q = 0$). FIGURE 17.23 compares an adiabatic process with isothermal and isochoric processes. We're now prepared to look at adiabatic processes in more detail.

In practice, there are two ways that an adiabatic process can come about. First, a gas cylinder can be completely surrounded by thermal insulation, such as thick pieces of Styrofoam. The environment can interact mechanically with the gas by pushing or pulling on the insulated piston, but there is no thermal interaction.

Second, the gas can be expanded or compressed very rapidly in what we call an *adiabatic expansion* or an *adiabatic compression*. In a rapid process there is essentially no time for heat to be transferred between the gas and the environment. We've already alluded to the idea that heat is transferred via atomic-level collisions. These collisions take time. If you stick one end of a copper rod into a flame, the other end will eventually get too hot to hold—but not instantly. Some amount of time is required for heat to be transferred from one end to the other. A process that takes place faster than the heat can be transferred is adiabatic.

NOTE ▶ You may recall reading in Chapter 16 that we are going to study only quasi-static processes, processes that proceed slowly enough to remain essentially in equilibrium at all times. Now we're proposing to study processes that take place very rapidly. Isn't this a contradiction? Yes, to some extent it is. What we need to establish are the appropriate time scales. How slow must a process go to be quasi-static? How fast must it go to be adiabatic? These types of calculations must be deferred to a more advanced course. It turns out—fortunately!—that many practical applications, such as the compression strokes in gasoline and diesel engines, are fast enough to be adiabatic yet slow enough to still be considered quasi-static. ◀

For an adiabatic process, with $Q = 0$, the first law of thermodynamics is $\Delta E_{th} = W$. Compressing a gas adiabatically ($W > 0$) increases the thermal energy. Thus **an adiabatic compression raises the temperature of a gas.** A gas that expands adiabatically ($W < 0$) gets colder as its thermal energy decreases. Thus **an adiabatic expansion lowers the temperature of a gas.** You can use an adiabatic process to change the gas temperature without using heat!

The work done in an adiabatic process goes entirely to changing the thermal energy of the gas. But we just found that $\Delta E_{th} = nC_V \Delta T$ for *any* process. Thus

$$W = nC_V \Delta T \qquad \text{(adiabatic process)} \qquad (17.34)$$

Equation 17.34 joins with the equations we derived earlier for the work done in isochoric, isobaric, and isothermal processes.

Gas processes can be represented as trajectories in the pV diagram. For example, a gas moves along a hyperbola during an isothermal process. How does an adiabatic process appear in a pV diagram? The result is more important than the derivation, which is a bit tedious, so we'll begin with the answer and then, at the end of this section, show where it comes from.

First, we define the **specific heat ratio** γ (lowercase Greek gamma) to be

$$\gamma = \frac{C_P}{C_V} = \begin{cases} 1.67 & \text{monatomic gas} \\ 1.40 & \text{diatomic gas} \end{cases} \qquad (17.35)$$

The specific heat ratio has many uses in thermodynamics. Notice that γ is dimensionless.

An adiabatic process is one in which

$$pV^\gamma = \text{constant} \qquad \text{or} \qquad p_f V_f^\gamma = p_i V_i^\gamma \qquad (17.36)$$

This is similar to the isothermal $pV = $ constant, but somewhat more complex due to the exponent γ.

The curves found by graphing $p = $ constant$/V^\gamma$ are called **adiabats.** In **FIGURE 17.24** you see that the two adiabats are steeper than the hyperbolic isotherms. An adiabatic process moves along an adiabat in the same way that an isothermal process moves along an isotherm. You can see that the temperature falls during an adiabatic expansion and rises during an adiabatic compression.

FIGURE 17.24 An adiabatic process moves along pV curves called *adiabats*.

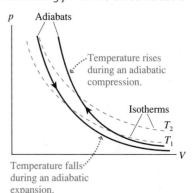

EXAMPLE 17.9 **An adiabatic compression**

Air containing gasoline vapor is admitted into the cylinder of an internal combustion engine at 1.00 atm pressure and 30°C. The piston rapidly compresses the gas from 500 cm³ to 50 cm³, a *compression ratio* of 10.

a. What are the final temperature and pressure of the gas?
b. Show the compression process on a pV diagram.
c. How much work is done to compress the gas?

MODEL The compression is rapid, with insufficient time for heat to be transferred from the gas to the environment, so we will model it as an adiabatic compression. We'll treat the gas as if it were 100% air.

SOLVE a. We know the initial pressure and volume, and we know the volume after the compression. For an adiabatic process, where pV^γ remains constant, the final pressure is

$$p_f = p_i \left(\frac{V_i}{V_f} \right)^\gamma = (1.00 \text{ atm})(10)^{1.40} = 25.1 \text{ atm}$$

Air is a mixture of N_2 and O_2, diatomic gases, so we used $\gamma = 1.40$. We can now find the temperature by using the ideal-gas law:

$$T_f = T_i \frac{p_f}{p_i} \frac{V_f}{V_i} = (303 \text{ K})(25.1)\left(\frac{1}{10} \right) = 761 \text{ K} = 488°C$$

Temperature *must* be in kelvins for doing gas calculations such as these.

b. **FIGURE 17.25** shows the pV diagram. The 30°C and 488°C isotherms are included to show how the temperature changes during the process.

FIGURE 17.25 The adiabatic compression of the gas in an internal combustion engine.

Continued

c. The work done is $W = nC_V \Delta T$, with $\Delta T = 458$ K. The number of moles is found from the ideal-gas law and the initial conditions:

$$n = \frac{p_i V_i}{RT_i} = 0.0201 \text{ mol}$$

Thus the work done to compress the gas is

$$W = nC_V \Delta T = (0.0201 \text{ mol})(20.8 \text{ J/mol K})(458 \text{ K}) = 192 \text{ J}$$

ASSESS The temperature rises dramatically during the compression stroke of an engine. But the higher temperature has nothing to do with heat! **The temperature and thermal energy of the gas are increased not by heating the gas but by doing work on it.** This is an important idea to understand.

If we use the ideal-gas-law expression $p = nRT/V$ in the adiabatic equation $pV^\gamma =$ constant, we see that $TV^{\gamma-1}$ is also constant during an adiabatic process. Thus another useful equation for adiabatic processes is

$$T_f V_f^{\gamma-1} = T_i V_i^{\gamma-1} \tag{17.37}$$

Proof of Equation 17.36

Now let's see where Equation 17.36 comes from. Consider an adiabatic process in which an infinitesimal amount of work dW done on a gas causes an infinitesimal change in the thermal energy. For an adiabatic process, with $dQ = 0$, the first law of thermodynamics is

$$dE_{th} = dW \tag{17.38}$$

We can use Equation 17.33, which is valid for *any* gas process, to write $dE_{th} = nC_V dT$. Earlier in the chapter we found that the work done during a small volume change is $dW = -p\, dV$. With these substitutions, Equation 17.38 becomes

$$nC_V dT = -p\, dV \tag{17.39}$$

The ideal-gas law can now be used to write $p = nRT/V$. The n cancels, and the C_V can be moved to the other side of the equation to give

$$\frac{dT}{T} = -\frac{R}{C_V}\frac{dV}{V} \tag{17.40}$$

We're going to integrate Equation 17.40, but anticipating the need for $\gamma = C_P/C_V$ we can first use the fact that $C_P = C_V + R$ to write

$$\frac{R}{C_V} = \frac{C_P - C_V}{C_V} = \frac{C_P}{C_V} - 1 = \gamma - 1 \tag{17.41}$$

Now we integrate Equation 17.40 from the initial state i to the final state f:

$$\int_{T_i}^{T_f} \frac{dT}{T} = -(\gamma - 1)\int_{V_i}^{V_f} \frac{dV}{V} \tag{17.42}$$

Carrying out the integration gives

$$\ln\left(\frac{T_f}{T_i}\right) = \ln\left(\frac{V_i}{V_f}\right)^{\gamma-1} \tag{17.43}$$

where we used the logarithm properties $\log a - \log b = \log(a/b)$ and $c \log a = \log(a^c)$. Taking the exponential of both sides now gives

$$\left(\frac{T_f}{T_i}\right) = \left(\frac{V_i}{V_f}\right)^{\gamma-1}$$
$$T_f V_f^{\gamma-1} = T_i V_i^{\gamma-1} \tag{17.44}$$

This was Equation 17.37. Writing $T = pV/nR$ and canceling $1/nR$ from both sides of the equation give Equation 17.36:

$$p_f V_f^\gamma = p_i V_i^\gamma \tag{17.45}$$

This was a lengthy derivation, but it is good practice at seeing how the ideal-gas law and the first law of thermodynamics can work together to yield results of great importance.

> **STOP TO THINK 17.6** For the two processes shown, which of the following is true:
>
> a. $Q_A > Q_B$
> b. $Q_A = Q_B$
> c. $Q_A < Q_B$

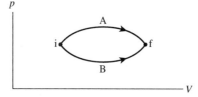

17.8 Heat-Transfer Mechanisms

You feel warmer when the sun is shining on you, colder when sitting on a metal bench or when the wind is blowing, especially if your skin is wet. This is due to the transfer of heat. Although we've talked about heat a lot in this chapter, we haven't said much about *how* heat is transferred from a hotter object to a colder object. There are four basic mechanisms by which objects exchange heat with their surroundings. Evaporation was treated in Section 17.5; in this section, we will consider the other mechanisms.

Heat-transfer mechanisms

When two objects are in direct contact, such as the soldering iron and the circuit board, heat is transferred by *conduction*.

Air currents near a warm glass of water rise, taking thermal energy with them in a process known as *convection*.

The lamp at the top shines on the lambs huddled below, warming them. The energy is transferred by *radiation*.

Blowing on a hot cup of tea or coffee cools it by *evaporation*.

Conduction

FIGURE 17.26 shows an object sandwiched between a higher temperature T_H and a lower temperature T_C. The temperature *difference* causes heat energy to be transferred from the hot side to the cold side in a process known as **conduction.**

It is not surprising that more heat is transferred if the temperature difference ΔT is larger. A material with a larger cross section A (a fatter pipe) transfers more heat, while a thicker material, increasing the distance L between the hot and cold sources, decreases the rate of heat transfer.

These observations about heat conduction can be summarized in a single formula. If heat Q is transferred in a time interval Δt, the *rate* of heat transfer is $Q/\Delta t$. For a material of cross-section area A and length L, spanning a temperature difference $\Delta T = T_H - T_C$, the rate of heat transfer is

$$\frac{Q}{\Delta t} = k\frac{A}{L}\Delta T \qquad (17.46)$$

FIGURE 17.26 Conduction of heat through a solid.

This material is conducting heat across the temperature difference.

Area A

T_H Hot

T_C Cold

Q

Q

L

TABLE 17.5 Thermal conductivities

Material	k (W/m K)
Diamond	2000
Silver	430
Copper	400
Aluminum	240
Iron	80
Stainless steel	14
Ice	1.7
Concrete	0.8
Glass	0.8
Styrofoam	0.035
Air (20°C, 1 atm)	0.023

The quantity k, which characterizes whether the material is a good conductor of heat or a poor conductor, is called the **thermal conductivity** of the material. Because the heat-transfer rate J/s is a *power,* measured in watts, the units of k are W/m K. Values of k for common materials are given in Table 17.5; a material with a larger value of k is a better conductor of heat.

Most good heat conductors are metals, which are also good conductors of electricity. One exception is diamond, in which the strong bonds among atoms that make diamond such a hard material lead to a rapid transfer of thermal energy. Air and other gases are poor conductors of heat because there are no bonds between adjacent molecules.

Some of our perceptions of hot and cold have more to do with thermal conductivity than with temperature. For example, a metal chair feels colder to your bare skin than a wooden chair not because it has a lower temperature—both are at room temperature—but because it has a much larger thermal conductivity that conducts heat away from your body at a much higher rate.

EXAMPLE 17.10 **Keeping a freezer cold**

A 1.8-m-wide by 1.0-m-tall by 0.65-m-deep home freezer is insulated with 5.0-cm-thick Styrofoam insulation. At what rate must the compressor remove heat from the freezer to keep the inside at −20°C in a room where the air temperature is 25°C?

MODEL Heat is transferred through each of the six sides by conduction. The compressor must remove heat at the same rate it enters to maintain a steady temperature inside. The heat conduction is determined primarily by the thick insulation, so we'll neglect the thin inner and outer panels.

SOLVE Each of the six sides is a slab of Styrofoam with cross-section area A_i and thickness $L = 5.0$ cm. The total rate of heat transfer is

$$\frac{Q}{\Delta t} = \sum_{i=1}^{6} k \frac{A_i}{L} \Delta T = \frac{k \Delta T}{L} \sum_{i=1}^{6} A_i = \frac{k \Delta T}{L} A_{total}$$

The total surface area is

$$A_{total} = 2 \times (1.8 \text{ m} \times 1.0 \text{ m} + 1.8 \text{ m} \times 0.65 \text{ m}$$
$$+ 1.0 \text{ m} \times 0.65 \text{ m}) = 7.24 \text{ m}^2$$

Using $k = 0.035$ W/m K from Table 17.5, we find

$$\frac{Q}{\Delta t} = \frac{k \Delta t}{L} A_{total} = \frac{(0.035 \text{ W/m K})(45 \text{ K})(7.24 \text{ m}^2)}{0.050 \text{ m}} = 230 \text{ W}$$

Heat enters the freezer through the walls at the rate 230 J/s; thus the compressor must remove 230 J of heat energy every second to keep the temperature at −20°C.

ASSESS We'll learn in Chapter 19 how the compressor does this and how much work it must do. A typical freezer uses electric energy at a rate of about 150 W, so our result seems reasonable.

Warm water (colored) moves by convection.

Convection

Air is a poor conductor of heat, but thermal energy is easily transferred through air, water, and other fluids because the air and water can flow. A pan of water on the stove is heated at the bottom. This heated water expands, becomes less dense than the water above it, and thus rises to the surface, while cooler, denser water sinks to take its place. The same thing happens to air. This transfer of thermal energy by the motion of a fluid—the well-known idea that "heat rises"—is called **convection.**

Convection is usually the main mechanism for heat transfer in fluid systems. On a small scale, convection mixes the pan of water that you heat on the stove; on a large scale, convection is responsible for making the wind blow and ocean currents circulate. Air is a very poor thermal conductor, but it is very effective at transferring energy by convection. To use air for thermal insulation, it is necessary to trap the air in small pockets to limit convection. And that's exactly what feathers, fur, double-paned windows, and fiberglass insulation do. Convection is much more rapid in water than in air, which is why people can die of hypothermia in 68°F (20°C) water but can live quite happily in 68°F air.

Because convection involves the often-turbulent motion of fluids, there is no simple equation for energy transfer by convection. Our description must remain qualitative.

Radiation

The sun *radiates* energy to earth through the vacuum of space. Similarly, you feel the warmth from the glowing red coals in a fireplace.

All objects emit energy in the form of **radiation,** electromagnetic waves generated by oscillating electric charges in the atoms that form the object. These waves transfer energy from the object that emits the radiation to the object that absorbs it. Electromagnetic waves carry energy from the sun; this energy is absorbed when sunlight falls on your skin, warming you by increasing your thermal energy. Your skin also emits electromagnetic radiation, helping to keep your body cool by decreasing your thermal energy. Radiation is a significant part of the *energy balance* that keeps your body at the proper temperature.

NOTE ▶ The word "radiation" comes from "radiate," meaning "to beam." Radiation can refer to x rays or to the radioactive decay of nuclei, but it also can refer simply to light and other forms of electromagnetic waves that "beam" from an object. Here we are using this second meaning of the term. ◀

You are familiar with radiation from objects hot enough to glow "red hot" or, at a high enough temperature, "white hot." The sun is simply a very hot ball of glowing gas, and the white light from an incandescent lightbulb is radiation emitted by a thin wire filament heated to a very high temperature by an electric current. Objects at lower temperatures also radiate, but at infrared wavelengths. You can't see this radiation (although you can sometimes feel it), but infrared-sensitive detectors can measure it and are used to make thermal images.

The energy radiated by an object depends strongly on temperature. If heat energy Q is radiated in a time interval Δt by an object with surface area A and absolute temperature T, the *rate* of heat transfer is found to be

$$\frac{Q}{\Delta t} = e\sigma A T^4 \qquad (17.47)$$

Because the rate of energy transfer is power (1 J/s = 1 W), $Q/\Delta t$ is often called the *radiated power.* Notice the very strong fourth-power dependence on temperature. Doubling the absolute temperature of an object increases the radiated power by a factor of 16!

The parameter e in Equation 17.47 is the **emissivity** of the surface, a measure of how effectively it radiates. The value of e ranges from 0 to 1. σ is a constant, known as the Stefan-Boltzmann constant, with the value

$$\sigma = 5.67 \times 10^{-8} \text{ W/m}^2 \text{ K}^4$$

NOTE ▶ Just as in the ideal-gas law, the temperature in Equation 17.47 *must* be in kelvins. ◀

Objects not only emit radiation, they also *absorb* radiation emitted by their surroundings. Suppose an object at temperature T is surrounded by an environment at temperature T_0. The *net* rate at which the object radiates heat energy—that is, radiation emitted minus radiation absorbed—is

$$\frac{Q_{\text{net}}}{\Delta t} = e\sigma A (T^4 - T_0^4) \qquad (17.48)$$

This makes sense. An object should have no *net* radiation if it's in thermal equilibrium ($T = T_0$) with its surroundings.

Notice that the emissivity e appears for absorption as well as emission; good emitters are also good absorbers. A perfect absorber ($e = 1$), one absorbing all light and radiation impinging on it but reflecting none, would appear completely black. Thus a perfect absorber is sometimes called a **black body.** But a perfect absorber would also be a perfect emitter, so thermal radiation from an ideal emitter is called **black-body radiation.** It seems strange that black objects are perfect emitters, but think of black charcoal glowing bright red in a fire. At room temperature, it "glows" equally bright with infrared.

This satellite image shows radiation emitted by the ocean waters off the east coast of the United States. You can clearly see the warm waters of the Gulf Stream, a large-scale convection that transfers heat to northern latitudes.

Taking the sun's temperature

The radius of the sun is 6.96×10^8 m. At the distance of the earth, 1.50×10^{11} m, the intensity of solar radiation (measured by satellites above the atmosphere) is 1370 W/m². What is the temperature of the sun's surface?

MODEL Assume the sun to be an ideal radiator with $e = 1$.

SOLVE The total power radiated by the sun is the power per m² multiplied by the surface area of a sphere extending to the earth:

$$P = \frac{1370 \text{ W}}{1 \text{ m}^2} \times 4\pi(1.50 \times 10^{11} \text{ m})^2 = 3.87 \times 10^{26} \text{ W}$$

That is, the sun radiates energy at the rate $Q/\Delta t = 3.87 \times 10^{26}$ J/s. That's a lot of power! This energy is radiated from the surface of a sphere of radius R_S. Using this information in Equation 17.47, we find that the sun's surface temperature is

$$T = \left[\frac{Q/\Delta t}{e\sigma(4\pi R_S^2)} \right]^{1/4}$$

$$= \left[\frac{3.87 \times 10^{26} \text{ W}}{(1)(5.67 \times 10^{-8} \text{ W/m}^2 \text{ K}^4)4\pi(6.96 \times 10^8 \text{ m})^2} \right]^{1/4}$$

$$= 5790 \text{ K}$$

ASSESS This temperature is confirmed by measurements of the solar spectrum, a topic we'll explore in Part VII.

Thermal radiation plays a prominent role in climate and global warming. The earth as a whole is in thermal equilibrium. Consequently, it must radiate back into space exactly as much energy as it receives from the sun. The incoming radiation from the hot sun is mostly visible light. The earth's atmosphere is transparent to visible light, so this radiation reaches the surface and is absorbed. The cooler earth radiates infrared radiation, but the atmosphere is *not* completely transparent to infrared. Some components of the atmosphere, notably water vapor and carbon dioxide, are strong absorbers of infrared radiation. They hinder the emission of radiation and, rather like a blanket, keep the earth's surface warmer than it would be without these gases in the atmosphere.

The **greenhouse effect,** as it's called, is a natural part of the earth's climate. The earth would be much colder and mostly frozen were it not for naturally occurring carbon dioxide in the atmosphere. But carbon dioxide also results from the burning of fossil fuels, and human activities since the beginning of the industrial revolution have increased the atmospheric concentration of carbon dioxide by nearly 50%. This human contribution has amplified the greenhouse effect and is the primary cause of global warming.

STOP TO THINK 17.7 Suppose you are an astronaut in space, hard at work in your sealed spacesuit. The only way that you can transfer excess heat to the environment is by

    a. Conduction.   b. Convection.   c. Radiation.   d. Evaporation.

Boiling water

400 mL of water is poured into an 8.0-cm-diameter, 150 g glass beaker with a 2.0-mm-thick bottom; then the beaker is placed on a 400°C hot plate. Once the water reaches the boiling point, how long will it take to boil away all the water?

MODEL The bottom of the beaker is a heat-conducting material transferring heat energy from the 400°C hot plate to the 100°C boiling water. The temperature of both the water and the beaker remains constant until the water has boiled away. We'll assume that heat losses due to convection and radiation are negligible, in which case the heat energy entering the system is used entirely for the phase change of the water. The beaker's mass isn't relevant because its temperature isn't changing.

SOLVE The heat energy required to boil mass M of water is

$$Q = ML_v$$

where $L_v = 2.26 \times 10^6$ J/kg is the heat of vaporization. The heat energy transferred through the bottom of the beaker during a time interval Δt is

$$Q = k\frac{A}{L}\Delta T \Delta t$$

where $k = 0.80$ W/m K is the thermal conductivity of glass. Because the heat transferred by conduction is used entirely for boiling the water, we can combine these two expressions:

$$k\frac{A}{L}\Delta T \Delta t = ML_v$$

and then solve for Δt:

$$\Delta t = \frac{MLL_v}{kA\Delta T} = \frac{(0.40 \text{ kg})(0.0020 \text{ m})(2.26 \times 10^6 \text{ J/kg})}{(0.80 \text{ W/m K})(0.0050 \text{ m}^2)(300 \text{ K})}$$

$$= 1500 \text{ s} = 25 \text{ min}$$

We used the density of water to find that $M = 400$ g $= 0.40$ kg and calculated $A = \pi r^2 = 0.0050$ m² as the area through which heat conduction occurs.

ASSESS 400 mL is roughly 2 cups, a small hot plate can bring 2 cups of water to a boil in 5 min or so, and boiling the water away takes quite a bit longer than bringing it to a boil. 25 min is probably a slight underestimate since we neglected energy losses due to convection and radiation, but it seems reasonable. A stove could boil the water away much faster because the burner temperature (gas flame or red-hot heating coil) is much higher.

SUMMARY

The goal of Chapter 17 has been to develop and apply the first law of thermodynamics.

General Principles

First Law of Thermodynamics

$$\Delta E_{th} = W + Q$$

The first law is a general statement of energy conservation.

Work W and heat Q depend on the process by which the system is changed.

The change in the system depends only on the total energy exchanged $W + Q$, not on the process.

Energy

Thermal energy E_{th} Microscopic energy of moving molecules and stretched molecular bonds. ΔE_{th} depends on the initial/final states but is independent of the process.

Work W Energy transferred to the system by forces in a mechanical interaction.

Heat Q Energy transferred to the system via atomic-level collisions when there is a temperature difference. A thermal interaction.

Important Concepts

The work done on a gas is

$$W = -\int_{V_i}^{V_f} p\,dV$$

$$= -(\text{area under the } pV \text{ curve})$$

An adiabatic process is one for which $Q = 0$. Gases move along an **adiabat** for which $pV^\gamma = $ constant, where $\gamma = C_P/C_V$ is the **specific heat ratio.** An adiabatic process changes the temperature of the gas without heating it.

Calorimetry When two or more systems interact thermally, they come to a common final temperature determined by

$$Q_{net} = Q_1 + Q_2 + \cdots = 0$$

The heat of transformation L is the energy needed to cause 1 kg of substance to undergo a phase change

$$Q = \pm ML$$

The specific heat c of a substance is the energy needed to raise the temperature of 1 kg by 1 K:

$$Q = Mc\Delta T$$

The molar specific heat C is the energy needed to raise the temperature of 1 mol by 1 K:

$$Q = nC\Delta T$$

The molar specific heat of gases depends on the *process* by which the temperature is changed:

$C_V = $ molar specific heat at **constant volume**

$C_P = C_V + R = $ molar specific heat at **constant pressure**

Heat is transferred by conduction, convection, radiation, and evaporation.

Conduction: $Q/\Delta t = (kA/L)\Delta T$

Radiation: $Q/\Delta t = e\sigma A T^4$

Summary of Basic Gas Processes

Process	Definition	Stays constant	Work	Heat
Isochoric	$\Delta V = 0$	V and p/T	$W = 0$	$Q = nC_V\Delta T$
Isobaric	$\Delta p = 0$	p and V/T	$W = -p\Delta V$	$Q = nC_P\Delta T$
Isothermal	$\Delta T = 0$	T and pV	$W = -nRT\ln(V_f/V_i)$	$\Delta E_{th} = 0$
Adiabatic	$Q = 0$	pV^γ	$W = \Delta E_{th}$	$Q = 0$
All gas processes	First law $\Delta E_{th} = W + Q = nC_V\Delta T$		Ideal-gas law $pV = nRT$	

Terms and Notation

work, W	adiabatic process	molar specific heat at	thermal conductivity, k
mechanical interaction	specific heat, c	constant volume, C_V	convection
mechanical equilibrium	molar specific heat, C	molar specific heat at	radiation
heat, Q	heat of transformation, L	constant pressure, C_P	emissivity, e
thermal interaction	heat of fusion, L_f	specific heat ratio, γ	black body
thermal equilibrium	heat of vaporization, L_v	adiabat	black-body radiation
first law of thermodynamics	calorimetry	conduction	greenhouse effect
thermodynamic energy model			

CONCEPTUAL QUESTIONS

1. When the space shuttle returns to earth, its surfaces get very hot as it passes through the atmosphere at high speed. Has the space shuttle been heated? If so, what was the source of the heat? If not, why is it hot?

2. Do (a) temperature, (b) heat, and (c) thermal energy describe a property of a system, an interaction of the system with its environment, or both? Explain.

3. Two containers hold equal masses of nitrogen gas at equal temperatures. You supply 10 J of heat to container A while not allowing its volume to change, and you supply 10 J of heat to container B while not allowing its pressure to change. Afterward, is temperature T_A greater than, less than, or equal to T_B? Explain.

4. You need to raise the temperature of a gas by 10°C. To use the least amount of heat energy, should you heat the gas at constant pressure or at constant volume? Explain.

5. *Why* is the molar specific heat of a gas at constant pressure larger than the molar specific heat at constant volume?

6. FIGURE Q17.6 shows an adiabatic process.
   a. Is the final temperature higher than, lower than, or equal to the initial temperature?
   b. Is any heat energy added to or removed from the system in this process? Explain.

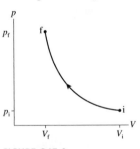

FIGURE Q17.6   FIGURE Q17.7

7. FIGURE Q17.7 shows two different processes taking an ideal gas from state i to state f. Is the work done on the gas in process A greater than, less than, or equal to the work done in process B? Explain.

8. FIGURE Q17.8 shows two different processes taking an ideal gas from state i to state f.
   a. Is the temperature *change* ΔT during process A larger than, smaller than, or equal to the change during process B? Explain.
   b. Is the heat energy added during process A greater than, less than, or equal to the heat added during process B? Explain.

FIGURE Q17.8   FIGURE Q17.9

9. Describe a series of steps in which you use the cylinder of Figure 17.13 to implement the ideal-gas process shown in FIGURE Q17.9. Then show the process as a first-law bar chart.

10. Describe a series of steps in which you use the cylinder of Figure 17.13 to implement the ideal-gas process shown in FIGURE Q17.10. Then show the process as a first-law bar chart.

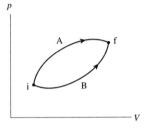

FIGURE Q17.10

11. The gas cylinder in FIGURE Q17.11, similar to the cylinder shown in Figure 17.13, is placed on a block of ice. The initial gas temperature is $> 0°C$.
    a. During the process that occurs until the gas reaches a new equilibrium, are (i) ΔT, (ii) W, and (iii) Q greater than, less than, or equal to zero? Explain.
    b. Draw a pV diagram showing the process.

FIGURE Q17.11

12. The gas cylinder in FIGURE Q17.12 is similar to the cylinder described earlier in Figure 17.13, except that the bottom is insulated. Masses are slowly removed from the top of the piston until the total mass is reduced by 50%.
    a. During this process, are (i) ΔT, (ii) W, and (iii) Q greater than, less than, or equal to zero? Explain.
    b. Draw a pV diagram showing the process.

FIGURE Q17.12

EXERCISES AND PROBLEMS

Problems labeled ▨ integrate material from earlier chapters.

Exercises

Section 17.1 It's All About Energy

Section 17.2 Work in Ideal-Gas Processes

1. ‖ How much work is done on the gas in the process shown in FIGURE EX17.1?

FIGURE EX17.1   FIGURE EX17.2

2. ‖ How much work is done on the gas in the process shown in FIGURE EX17.2?

3. ‖ 80 J of work are done on the gas in the process shown in FIGURE EX17.3. What is V_1 in cm³?

FIGURE EX17.3

4. ‖ A 2000 cm³ container holds 0.10 mol of helium gas at 300°C. How much work must be done to compress the gas to 1000 cm³ at (a) constant pressure and (b) constant temperature?

Section 17.3 Heat

Section 17.4 The First Law of Thermodynamics

5. | Draw a first-law bar chart (see Figure 17.14) for the gas process in FIGURE EX17.5.

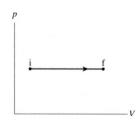

FIGURE EX17.5

6. | Draw a first-law bar chart (see Figure 17.14) for the gas process in FIGURE EX17.6.

FIGURE EX17.6

7. | Draw a first-law bar chart (see Figure 17.14) for the gas process in FIGURE EX17.7.

 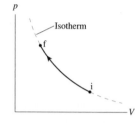

FIGURE EX17.7   FIGURE EX17.8

8. | Draw a first-law bar chart (see Figure 17.14) for the gas process in FIGURE EX17.8.

9. ‖ A gas is compressed from 600 cm³ to 200 cm³ at a constant pressure of 400 kPa. At the same time, 100 J of heat energy is transferred out of the gas. What is the change in thermal energy of the gas during this process?

10. | 500 J of work are done on a system in a process that decreases the system's thermal energy by 200 J. How much heat energy is transferred to or from the system?

Section 17.5 Thermal Properties of Matter

11. ‖ How much heat energy must be added to a 6.0 cm × 6.0 cm × 6.0 cm block of aluminum to raise its temperature from − 50°C to 50°C?

12. ‖ A rapidly spinning paddle wheel raises the temperature of 200 mL of water from 21°C to 25°C. How much (a) heat is transferred and (b) work is done in this process?

13. | a. 100 J of heat energy are transferred to 20 g of mercury. By how much does the temperature increase?
    b. How much heat is needed to raise the temperature of 20 g of water by the same amount?

14. ‖ How much heat is needed to change 20 g of mercury at 20°C into mercury vapor at the boiling point?

15. ‖ What is the maximum mass of ethyl alcohol you could boil with 1000 J of heat, starting from 20°C?

Section 17.6 Calorimetry

16. ‖ 30 g of copper pellets are removed from a 300°C oven and immediately dropped into 100 mL of water at 20°C in an insulated cup. What will the new water temperature be?

17. ‖ A 750 g aluminum pan is removed from the stove and plunged into a sink filled with 10.0 L of water at 20.0°C. The water temperature quickly rises to 24.0°C. What was the initial temperature of the pan in °C and in °F?

18. ‖ A 50.0 g thermometer is used to measure the temperature of 200 mL of water. The specific heat of the thermometer, which is mostly glass, is 750 J/kg K, and it reads 20.0°C while lying on the table. After being completely immersed in the water, the thermometer's reading stabilizes at 71.2°C. What was the actual water temperature before it was measured?

19. ‖ A 500 g metal sphere is heated to 300°C, then dropped into a beaker containing 300 cm³ of mercury at 20.0°C. A short time later the mercury temperature stabilizes at 99.0°C. Identify the metal.

20. ‖ A 65 cm³ block of iron is removed from an 800°C furnace and immediately dropped into 200 mL of 20°C water. What fraction of the water boils away?

Section 17.7 The Specific Heats of Gases

21. ‖ A container holds 1.0 g of argon at a pressure of 8.0 atm.
    a. How much heat is required to increase the temperature by 100°C at constant volume?
    b. How much will the temperature increase if this amount of heat energy is transferred to the gas at constant pressure?

22. ‖ A container holds 1.0 g of oxygen at a pressure of 8.0 atm.
    a. How much heat is required to increase the temperature by 100°C at constant pressure?
    b. How much will the temperature increase if this amount of heat energy is transferred to the gas at constant volume?

23. ‖ A rigid cylinder contains 7.0 g of nitrogen at 20°C. What is the minimum amount of heat energy that must be removed to liquify the nitrogen?

24. ‖ The volume of a gas is halved during an adiabatic compression that increases the pressure by a factor of 2.5.
    a. What is the specific heat ratio γ?
    b. By what factor does the temperature increase?

25. ‖ A gas cylinder holds 0.10 mol of O_2 at 150°C and a pressure of 3.0 atm. The gas expands adiabatically until the pressure is halved. What are the final (a) volume and (b) temperature?

26. ‖ A gas cylinder holds 0.10 mol of O_2 at 150°C and a pressure of 3.0 atm. The gas expands adiabatically until the volume is doubled. What are the final (a) pressure and (b) temperature?

Section 17.8 Heat-Transfer Mechanisms

27. ‖ A 10 m × 14 m house is built on a 12-cm-thick concrete slab. What is the heat-loss rate through the slab if the ground temperature is 5°C while the interior of the house is 22°C?

28. ‖ The ends of a 20-cm-long, 2.0-cm-diameter rod are maintained at 0°C and 100°C by immersion in an ice-water bath and boiling water. Heat is conducted through the rod at 4.5×10^4 J per hour. Of what material is the rod made?

29. ‖ What maximum power can be radiated by a 10-cm-diameter solid lead sphere? Assume an emissivity of 1.

30. ‖ BIO Radiation from the head is a major source of heat loss from the human body. Model a head as a 20-cm-diameter, 20-cm-tall cylinder with a flat top. If the body's surface temperature is 35°C, what is the net rate of heat loss on a chilly 5°C day? All skin, regardless of color, is effectively black in the infrared where the radiation occurs, so use an emissivity of 0.95.

Problems

31. ‖ A 5.0 g ice cube at −20°C is in a rigid, sealed container from which all the air has been evacuated. How much heat is required to change this ice cube into steam at 200°C?

32. ‖ A 5.0-m-diameter garden pond is 30 cm deep. Solar energy is incident on the pond at an average rate of 400 W/m². If the water absorbs all the solar energy and does not exchange energy with its surroundings, how many hours will it take to warm from 15°C to 25°C?

33. ‖ An 11 kg bowling ball at 0°C is dropped into a tub containing a mixture of ice and water. A short time later, when a new equilibrium has been established, there are 5.0 g less ice. From what height was the ball dropped? Assume no water or ice splashes out.

34. ‖ The burner on an electric stove has a power output of 2.0 kW. A 750 g stainless steel teakettle is filled with 20°C water and placed on the already hot burner. If it takes 3.0 min for the water to reach a boil, what volume of water, in cm³, was in the kettle? Stainless steel is mostly iron, so you can assume its specific heat is that of iron.

35. ‖ BIO Reptiles don't use enough metabolic energy to keep their body temperature constant. They cool off at night and must warm up in the morning sun. Suppose a 2.9-m-long, 60-cm-wide, 350 kg alligator is basking in the sun. If the sun's intensity on the back of the alligator is 500 W/m², and if energy losses can be ignored, how long will it take the alligator to warm up from 23°C to a more favorable 30°C? The average specific heat of body tissue is 3400 J/kg K.

36. ‖ BIO One way you keep from overheating is by perspiring. Evaporation—a phase change—requires heat, and the heat energy is removed from your body. Evaporation is much like boiling, only water's heat of vaporization at 35°C is a somewhat larger 24×10^5 J/kg because at lower temperatures more energy is required to break the molecular bonds. Very strenuous activity can cause an adult human to produce 30 g of perspiration per minute. If all the perspiration evaporates, rather than dripping off, at what rate (in J/s) is it possible to exhaust heat by perspiring?

37. ‖ BIO When air is inhaled, it quickly becomes saturated with water vapor as it passes through the moist airways. Consequently, an adult human exhales about 25 mg of evaporated water with each breath. Evaporation—a phase change—requires heat, and the heat energy is removed from your body. Evaporation is much like boiling, only water's heat of vaporization at 35°C is a somewhat larger 24×10^5 J/kg because at lower temperatures more energy is required to break the molecular bonds. At 12 breaths/min, on a dry day when the inhaled air has almost no water content, what is the body's rate of energy loss (in J/s) due to exhaled water? (For comparison, the energy loss from radiation, usually the largest loss on a cool day, is about 100 J/s.)

38. ‖ Two cars collide head-on while each is traveling at 80 km/h. Suppose all their kinetic energy is transformed into the thermal energy of the wrecks. What is the temperature increase of each car? You can assume that each car's specific heat is that of iron.

39. ‖ 10 g of aluminum at 200°C and 20 g of copper are dropped into 50 cm³ of ethyl alcohol at 15°C. The temperature quickly comes to 25°C. What was the initial temperature of the copper?

40. ‖ A 100 g ice cube at −10°C is placed in an aluminum bucket whose initial temperature is 70°C. The system comes to an equilibrium temperature of 20°C. What is the mass of the bucket?

41. ‖ 512 g of an unknown metal at a temperature of 15°C is dropped into a 100 g aluminum container holding 325 g of water at 98°C. A short time later, the container of water and metal stabilizes at a new temperature of 78°C. Identify the metal.

42. ‖ A 150 L ($\approx$ 40 gal) electric hot-water tank has a 5.0 kW heater. How many minutes will it take to raise the water temperature from 65°F to 140°F?

43. ‖ What is oxygen's specific heat at constant volume in J/kg K?

44. ‖ BIO Suppose you take and hold a deep breath on a chilly day, inhaling 3.0 L of air at 0°C and 1 atm.
    a. How much heat must your body supply to warm the air to your internal body temperature of 37°C?
    b. By how much does the air's volume increase as it warms?

45. | An experiment measures the temperature of a 500 g substance while steadily supplying heat to it. **FIGURE P17.45** shows the results of the experiment. What are the (a) specific heat of the solid phase, (b) specific heat of the liquid phase, (c) melting and boiling temperatures, and (d) heats of fusion and vaporization?

FIGURE P17.45

46. ‖ Your 300 mL cup of coffee is too hot to drink when served at 90°C. What is the mass of an ice cube, taken from a −20°C freezer, that will cool your coffee to a pleasant 60°C?

47. ‖ A typical nuclear reactor generates 1000 MW (1000 MJ/s) of electrical energy. In doing so, it produces 2000 MW of "waste heat" that must be removed from the reactor to keep it from melting down. Many reactors are sited next to large bodies of water so that they can use the water for cooling. Consider a reactor where the intake water is at 18°C. State regulations limit the temperature of the output water to 30°C so as not to harm aquatic organisms. How many liters of cooling water have to be pumped through the reactor each minute?

48. ‖ A beaker with a metal bottom is filled with 20 g of water at 20°C. It is brought into good thermal contact with a 4000 cm³ container holding 0.40 mol of a monatomic gas at 10 atm pressure. Both containers are well insulated from their surroundings. What is the gas pressure after a long time has elapsed? You can assume that the containers themselves are nearly massless and do not affect the outcome.

FIGURE P17.48

49. ‖ 2.0 mol of gas are at 30°C and a pressure of 1.5 atm. How much work must be done on the gas to compress it to one third of its initial volume at (a) constant temperature and (b) constant pressure? (c) Show both processes on a single pV diagram.

50. ‖ 500 J of work must be done to compress a gas to half its initial volume at constant temperature. How much work must be done to compress the gas by a factor of 10, starting from its initial volume?

51. ‖ A 6.0-cm-diameter cylinder of nitrogen gas has a 4.0-cm-thick movable copper piston. The cylinder is oriented vertically, as shown in **FIGURE P17.51**, and the air above the piston is evacuated. When the gas temperature is 20°C, the piston floats 20 cm above the bottom of the cylinder.
   a. What is the gas pressure?
   b. How many gas molecules are in the cylinder?
   Then 2.0 J of heat energy are transferred to the gas.
   c. What is the new equilibrium temperature of the gas?
   d. What is the final height of the piston?
   e. How much work is done on the gas as the piston rises?

FIGURE P17.51

52. ‖ A 10-cm-diameter cylinder contains argon gas at 10 atm pressure and a temperature of 50°C. A piston can slide in and out of the cylinder. The cylinder's initial length is 20 cm. 2500 J of heat are transferred to the gas, causing the gas to expand at constant pressure. What are (a) the final temperature and (b) the final length of the cylinder?

53. ‖ A cube 20 cm on each side contains 3.0 g of helium at 20°C. 1000 J of heat energy are transferred to this gas. What are (a) the final pressure if the process is at constant volume and (b) the final volume if the process is at constant pressure? (c) Show and label both processes on a single pV diagram.

54. ‖ An 8.0-cm-diameter, well-insulated vertical cylinder containing nitrogen gas is sealed at the top by a 5.1 kg frictionless piston. The air pressure above the piston is 100 kPa.
   a. What is the gas pressure inside the cylinder?
   b. Initially, the piston height above the bottom of the cylinder is 26 cm. What will be the piston height if an additional 3.5 kg are placed on top of the piston?

55. ‖ n moles of an ideal gas at temperature T_1 and volume V_1 expand isothermally until the volume has doubled. In terms of n, T_1, and V_1, what are (a) the final temperature, (b) the work done on the gas, and (c) the heat energy transferred to the gas?

56. ‖ 5.0 g of nitrogen gas at 20°C and an initial pressure of 3.0 atm undergo an isobaric expansion until the volume has tripled.
   a. What are the gas volume and temperature after the expansion?
   b. How much heat energy is transferred to the gas to cause this expansion?
   The gas pressure is then decreased at constant volume until the original temperature is reached.
   c. What is the gas pressure after the decrease?
   d. What amount of heat energy is transferred from the gas as its pressure decreases?
   e. Show the total process on a pV diagram. Provide an appropriate scale on both axes.

57. ‖ **FIGURE P17.57** shows two processes that take a gas from state i to state f. Show that $Q_A - Q_B = p_iV_i$.

FIGURE P17.57   **FIGURE P17.58**

58. | 0.10 mol of nitrogen gas follow the two processes shown in **FIGURE P17.58**. How much heat is required for each?

59. ‖ 0.10 mol of nitrogen gas follow the two processes shown in **FIGURE P17.59**. How much heat is required for each?

FIGURE P17.59

60. ‖ 0.10 mol of a monatomic gas follow the process shown in FIGURE P17.60.
    a. How much heat energy is transferred to or from the gas during process $1 \rightarrow 2$?
    b. How much heat energy is transferred to or from the gas during process $2 \rightarrow 3$?
    c. What is the total change in thermal energy of the gas?

FIGURE P17.60

61. ‖‖ Your laboratory assignment for the week is to measure the specific heat ratio γ of carbon dioxide. The gas is contained in a cylinder with a movable piston and a thermometer. When the piston is withdrawn as far as possible, the cylinder's length is 20 cm. You decide to push the piston in very rapidly by various amounts and, for each push, to measure the temperature of the carbon dioxide. Before each push, you withdraw the piston all the way and wait several minutes for the gas to come to the room temperature of 21°C. Your data are as follows:

Push (cm)	Temperature (°C)
5	35
10	68
13	110
15	150

Use the best-fit line of an appropriate graph to determine γ for carbon dioxide.

62. ‖ Two cylinders each contain 0.10 mol of a diatomic gas at 300 K and a pressure of 3.0 atm. Cylinder A expands isothermally and cylinder B expands adiabatically until the pressure of each is 1.0 atm.
    a. What are the final temperature and volume of each?
    b. Show both processes on a single pV diagram. Use an appropriate scale on both axes.

63. ‖‖ A monatomic gas follows the process $1 \rightarrow 2 \rightarrow 3$ shown in FIGURE P17.63. How much heat is needed for (a) process $1 \rightarrow 2$ and (b) process $2 \rightarrow 3$?

FIGURE P17.63

FIGURE P17.64

64. ‖ FIGURE P17.64 shows a thermodynamic process followed by 0.015 mol of hydrogen. How much heat energy is transferred to the gas?

65. ‖ FIGURE P17.65 shows a thermodynamic process followed by 120 mg of helium.
    a. Determine the pressure (in atm), temperature (in °C), and volume (in cm^3) of the gas at points 1, 2, and 3. Put your results in a table for easy reading.
    b. How much work is done on the gas during each of the three segments?
    c. How much heat energy is transferred to or from the gas during each of the three segments?

FIGURE P17.65

66. ‖ a. What compression ratio V_{max}/V_{min} will raise the air temperature from 20°C to 1000°C in an adiabatic process?
    b. What pressure ratio p_{max}/p_{min} does this process have?

67. ‖ Two containers of a diatomic gas have the same initial conditions. One container, heated at constant pressure, has a temperature increase of 20°C. The other container receives the same quantity of heat energy, but at constant volume. What is its temperature increase?

68. ‖ 14 g of nitrogen gas at STP are adiabatically compressed to a pressure of 20 atm. What are (a) the final temperature, (b) the work done on the gas, (c) the heat input to the gas, and (d) the compression ratio V_{max}/V_{min}? (e) Show the process on a pV diagram, using proper scales on both axes.

69. ‖ 14 g of nitrogen gas at STP are pressurized in an isochoric process to a pressure of 20 atm. What are (a) the final temperature, (b) the work done on the gas, (c) the heat input to the gas, and (d) the pressure ratio p_{max}/p_{min}? (e) Show the process on a pV diagram, using proper scales on both axes.

70. ‖ When strong winds rapidly carry air down from mountains to a lower elevation, the air has no time to exchange heat with its surroundings. The air is compressed as the pressure rises, and its temperature can increase dramatically. These warm winds are called Chinook winds in the Rocky Mountains and Santa Ana winds in California. Suppose the air temperature high in the mountains behind Los Angeles is 0°C at an elevation where the air pressure is 60 kPa. What will the air temperature be, in °C and °F, when the Santa Ana winds have carried this air down to an elevation near sea level where the air pressure is 100 kPa?

71. ‖ You would like to put a solar hot water system on your roof, but you're not sure it's feasible. A reference book on solar energy shows that the ground-level solar intensity in your city is 800 W/m² for at least 5 hours a day throughout most of the year. Assuming that a completely black collector plate loses energy only by radiation, and that the air temperature is 20°C, what is the equilibrium temperature of a collector plate directly facing the sun? Note that while a plate has two sides, only the side facing the sun will radiate because the opposite side will be well insulated.

72. ‖ A cubical box 20 cm on a side is constructed from 1.2-cm-thick concrete panels. A 100 W lightbulb is sealed inside the box. What is the air temperature inside the box when the light is on if the surrounding air temperature is 20°C?

73. ‖ The sun's intensity at the distance of the earth is 1370 W/m². 30% of this energy is reflected by water and clouds; 70% is absorbed. What would be the earth's average temperature (in °C) if the earth had no atmosphere? The emissivity of the surface is very close to 1. (The actual average temperature of the earth, about 15°C, is higher than your calculation because of the greenhouse effect.)

In Problems 74 through 76 you are given the equation used to solve a problem. For each of these, you are to
  a. Write a realistic problem for which this is the correct equation.
  b. Finish the solution of the problem.

74. $50 \text{ J} = -n(8.31 \text{ J/mol K})(350 \text{ K})\ln\left(\frac{1}{3}\right)$

75. $(200 \times 10^{-6} \text{ m}^3)(13{,}600 \text{ kg/m}^3)$
$\times (140 \text{ J/kg K})(90°C - 15°C)$
$+ (0.50 \text{ kg})(449 \text{ J/kg K})(90°C - T_i) = 0$

76. $(10 \text{ atm})V_2^{1.40} = (1.0 \text{ atm})V_1^{1.40}$

Challenge Problems

77. **FIGURE CP17.77** shows a thermodynamic process followed by 120 mg of helium.
  a. Determine the pressure (in atm), temperature (in °C), and volume (in cm³) of the gas at points 1, 2, and 3. Put your results in a table for easy reading.

FIGURE CP17.77

  b. How much work is done on the gas during each of the three segments?
  c. How much heat is transferred to or from the gas during each of the three segments?

78. One cylinder in the diesel engine of a truck has an initial volume of 600 cm³. Air is admitted to the cylinder at 30°C and a pressure of 1.0 atm. The piston rod then does 400 J of work to rapidly compress the air. What are its final temperature and volume?

79. You come into lab one day and find a well-insulated 2000 mL thermos bottle containing 500 mL of boiling liquid nitrogen. The remainder of the thermos has nitrogen gas at a pressure of 1.0 atm. The gas and liquid are in thermal equilibrium. While waiting for lab to start, you notice a piece of iron on the table with "197 g" written on it. Just for fun, you drop the iron into the thermos and seal the cap tightly so that no gas can escape. After a few seconds have passed, what is the pressure inside the thermos? The density of liquid nitrogen is 810 kg/m³.

80. A cylindrical copper rod and an iron rod with exactly the same dimensions are welded together end to end. The outside end of the copper rod is held at 100°C, and the outside end of the iron rod is held at 0°C. What is the temperature at the midpoint where the rods are joined together?

81. 0.020 mol of a diatomic gas, with initial temperature 20°C, are compressed from 1500 cm³ to 500 cm³ in a process in which $pV^2 = $ constant. How much heat is added during this process?

82. A monatomic gas fills the left end of the cylinder in **FIGURE CP17.82**. At 300 K, the gas cylinder length is 10.0 cm and the spring is compressed by 2.0 cm. How much heat energy must be added to the gas to expand the cylinder length to 16.0 cm?

FIGURE CP17.82

STOP TO THINK ANSWERS

Stop to Think 17.1: a. The piston does work W on the gas. There's no heat because of the insulation, and $\Delta E_{mech} = 0$ because the gas as a whole doesn't move. Thus $\Delta E_{th} = W > 0$. The work increases the system's thermal energy and thus raises its temperature.

Stop to Think 17.2: d. $W_A = 0$ because A is an isochoric process. $W_B = W_{1 \text{ to } 2} + W_{2 \text{ to } 3}$. $|W_{2 \text{ to } 3}| > |W_{1 \text{ to } 2}|$ because there's more area under the curve, and $W_{2 \text{ to } 3}$ is positive whereas $W_{1 \text{ to } 2}$ is negative. Thus W_B is positive.

Stop to Think 17.3: b and e. The temperature rises in d from doing work on the gas ($\Delta E_{th} = W$), not from heat. e involves heat because there is a temperature difference. The temperature of the gas doesn't change because the heat is used to do the work of lifting a weight.

Stop to Think 17.4: c. The temperature increases so E_{th} must increase. W is negative in an expansion, so Q must be positive and larger than $|W|$.

Stop to Think 17.5: a. A has a smaller specific heat and thus less thermal inertia. The temperature of A will change more than the temperature of B.

Stop to Think 17.6: a. $W_A + Q_A = W_B + Q_B$. The area under process A is larger than the area under B, so W_A is *more negative* than W_B. Q_A has to be more positive than Q_B to maintain the equality.

Stop to Think 17.7: c. Conduction, convection, and evaporation require matter. Only radiation transfers energy through the vacuum of space.

18 The Micro/Macro Connection

Heating the air in a hot-air balloon increases the thermal energy of the molecules. This causes the gas to expand, lowering its density and allowing it to float in the cooler surrounding air.

▶ **Looking Ahead** The goal of Chapter 18 is to understand a macroscopic system in terms of the microscopic behavior of its molecules.

Macro Puzzles Micro Explanations

- Why does the ideal-gas law work for every gas?
- Why is the molar specific heat for every monatomic gas the same? And for every diatomic gas and for every elemental solid?
- What does temperature measure?

These are puzzles we uncovered in the last two chapters. In this chapter, you will learn that we can resolve these puzzles and understand many of the properties of macroscopic systems by investigating the microscopic behavior of its atoms and molecules.

Macro	**Micro**
A container of an ideal gas	N molecules of gas with number density N/V

Macroscopic properties include pressure, temperature and thermal energy.

The molecules have kinetic energy. They collide with each other and the walls of the container.

Collisions

You will learn how to understand the pressure of a gas in terms of atomic collisions with the walls of the container.

Hotter   Colder

Q

You'll also learn that heat transfer occurs via collisions at the boundary between two systems. More-energetic molecules on one side transfer their energy to less-energetic molecules on the other.

Energy and Temperature

We'll find that the average energy of a molecule depends only on temperature.

This will allow us to interpret temperature in terms of thermal energy—the microscopic energy of the moving molecules—and then to understand why every monatomic gas has the same molar specific heat.

◀ **Looking Back**
Sections 17.3–17.7 Heat, the first law of thermodynamics, and specific heat

The Second Law of Thermodynamics

You will learn a new law of nature, the second law of thermodynamics, that governs how systems evolve in time.

One statement of the second law is that heat energy is transferred spontaneously from a hotter system to a colder system, never from colder to hotter. Heat transfer is an *irreversible process.*

Heat goes from the burner to the teakettle, making the water hotter and the burner a little cooler. It would not violate energy conservation for heat to go from the kettle to the burner, making the water colder and the burner hotter. But it doesn't happen.

You'll learn to use the concept of **entropy** to understand why all macroscopic interactions are irreversible. Entropy explains why the future is different from the past and why there are theoretical limits to the efficiency of using energy in practical ways.

18.1 Molecular Speeds and Collisions

Let us begin by thinking about gases at the atomic level. If gases really are composed of atoms and molecules in motion, how fast are the molecules moving? Do all molecules move with the same speed, or is there a range of speeds?

To answer these questions, FIGURE 18.1 shows an experiment to measure the speeds of molecules in a gas. The two rotating disks form a *velocity selector.* Once every revolution, the slot in the first disk allows a small pulse of molecules to pass through. By the time these molecules reach the second disk, the slots have rotated. The molecules can pass through the second slot and be detected *only* if they have exactly the right speed $v = L/\Delta t$ to travel between the two disks during time interval Δt it takes the axle to complete one revolution. Molecules having any other speed are blocked by the second disk. By changing the rotation period of the axle, this apparatus can measure how many molecules have each of many possible speeds.

FIGURE 18.2 shows the results for nitrogen gas (N_2) at $T = 20°C$. The data are presented in the form of a **histogram,** a bar chart in which the height of each bar tells how many (or, in this case, what percentage) of the molecules have a speed in the *range* of speeds shown below the bar. For example, 16% of the molecules have speeds in the range from 600 m/s to 700 m/s. All the bars sum to 100%, showing that this histogram describes *all* of the molecules leaving the source.

It turns out that the molecules have what is called a *distribution* of speeds, ranging from as low as $\approx$ 100 m/s to as high as $\approx$ 1200 m/s. But not all speeds are equally likely; there is a *most likely speed* of $\approx$ 550 m/s. This is really fast, $\approx$ 1200 mph! Changing the temperature or changing to a different gas changes the most likely speed, as we'll learn later in the chapter, but it does not change the *shape* of the distribution.

If you were to repeat the experiment, you would again find the most likely speed to be $\approx$ 550 m/s and that 16% of the molecules have speeds between 600 m/s and 700 m/s. Think about what this means. The "molecular deck of cards" is constantly being reshuffled by molecular collisions, yet 16% of the molecules always have speeds between 600 m/s and 700 m/s.

This is an important lesson. Although a gas consists of a vast number of molecules, each moving randomly, *averages,* such as the average number of molecules in the speed range 600 to 700 m/s, have precise, predictable values. **The micro/macro connection is built on the idea that the macroscopic properties of a system, such as temperature or pressure, are related to the *average* behavior of the atoms and molecules.**

Mean Free Path

Imagine someone opening a bottle of strong perfume a few feet away from you. If molecular speeds are hundreds of meters per second, you might expect to smell the perfume almost instantly. But that isn't what happens. As you know, it takes many seconds for the molecules to *diffuse* across the room. Let's see why this is.

FIGURE 18.3 shows a "movie" of one molecule. Instead of zipping along in a straight line, as it would in a vacuum, the molecule follows a convoluted zig-zag path in which it frequently collides with other molecules. A molecule may have traveled hundreds of meters by the time it manages to get 1 or 2 m away from its starting point.

The random distribution of the molecules in the gas causes the straight-line segments between collisions to be of unequal lengths. A question we could ask is: What is the *average* distance between collisions? If a molecule has N_{coll} collisions as it travels distance L, the average distance between collisions, which is called the **mean free path** λ (lowercase Greek lambda), is

$$\lambda = \frac{L}{N_{coll}} \qquad (18.1)$$

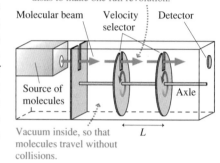

FIGURE 18.1 An experiment to measure the speeds of molecules in a gas.

The only molecules that reach the detector are those whose speed allows them to travel distance L during the time it takes the disks to make one full revolution.

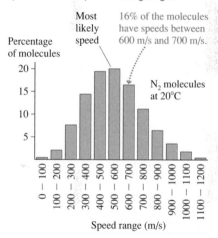

FIGURE 18.2 The distribution of molecular speeds in a sample of nitrogen gas.

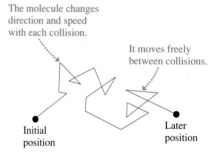

FIGURE 18.3 A single molecule follows a zig-zag path through a gas as it collides with other molecules.

The molecule changes direction and speed with each collision.

It moves freely between collisions.

Initial position

Later position

FIGURE 18.4 A sample molecule will collide with all target molecules whose centers are within a bent cylinder of radius $2r$ centered on its path.

(a)   Two molecules will collide if the distance between their centers is less than $2r$.

$2r$

$2r$

Molecules of radius r

(b)   Target molecules

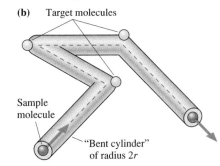

Sample molecule

"Bent cylinder" of radius $2r$

The concept of mean free path is used not only in gases but also to describe electrons moving through conductors and light passing through a medium that scatters the photons.

Our task is to determine the number of collisions. FIGURE 18.4a shows two molecules approaching each other. We will assume that the molecules are spherical and of radius r. We will also continue the ideal-gas assumption that the molecules undergo hard-sphere collisions, like billiard balls. In that case, the molecules will collide if the distance between their *centers* is less than $2r$. They will miss if the distance is greater than $2r$.

FIGURE 18.4b shows a cylinder of radius $2r$ centered on the trajectory of a "sample" molecule. The sample molecule collides with any "target" molecule whose center is located within the cylinder, causing the cylinder to bend at that point. Hence the number of collisions N_{coll} is equal to the number of molecules in a cylindrical volume of length L.

The volume of a cylinder is $V_{cyl} = AL = \pi(2r)^2 L$. If the number density of the gas is N/V particles per m^3, then the number of collisions along a trajectory of length L is

$$N_{coll} = \frac{N}{V} V_{cyl} = \frac{N}{V} \pi(2r)^2 L = 4\pi \frac{N}{V} r^2 L \qquad (18.2)$$

Thus the mean free path between collisions is

$$\lambda = \frac{L}{N_{coll}} = \frac{1}{4\pi(N/V)r^2}$$

We made a tacit assumption in this derivation that the target molecules are at rest. While the general idea behind our analysis is correct, a more detailed calculation with all the molecules moving introduces an extra factor of $\sqrt{2}$, giving

$$\lambda = \frac{1}{4\sqrt{2}\,\pi(N/V)r^2} \qquad \text{(mean free path)} \qquad (18.3)$$

Laboratory measurements are necessary to determine atomic and molecular radii, but a reasonable rule of thumb is to assume that atoms in a monatomic gas have $r \approx 0.5 \times 10^{-10}$ m and diatomic molecules have $r \approx 1.0 \times 10^{-10}$ m.

EXAMPLE 18.1 **The mean free path at room temperature**

What is the mean free path of a nitrogen molecule at 1.0 atm pressure and room temperature (20°C)?

SOLVE Nitrogen is a diatomic molecule, so $r \approx 1.0 \times 10^{-10}$ m. We can use the ideal-gas law in the form $pV = Nk_BT$ to determine the number density:

$$\frac{N}{V} = \frac{p}{k_BT} = \frac{101{,}300 \text{ Pa}}{(1.38 \times 10^{-23} \text{ J/K})(293 \text{ K})} = 2.5 \times 10^{25} \text{ m}^{-3}$$

Thus the mean free path is

$$\lambda = \frac{1}{4\sqrt{2}\,\pi(N/V)r^2}$$
$$= \frac{1}{4\sqrt{2}\,\pi(2.5 \times 10^{25} \text{ m}^{-3})(1.0 \times 10^{-10} \text{ m})^2}$$
$$= 2.3 \times 10^{-7} \text{ m} = 230 \text{ nm}$$

ASSESS You learned in Example 16.5 that the average separation between gas molecules at STP is ≈ 4 nm. It seems that any given molecule can slip between its neighbors, which are spread out in three dimensions, and travel—on average—about 60 times the average spacing before it collides with another molecule.

STOP TO THINK 18.1   The table shows the properties of four gases, each having the same number of molecules. Rank in order, from largest to smallest, the mean free paths λ_A to λ_D of molecules in these gases.

Gas	A	B	C	D
Volume	V	$2V$	V	V
Atomic mass	m	m	$2m$	m
Atomic radius	r	r	r	$2r$

18.2 Pressure in a Gas

Why does a gas have pressure? In Chapter 15, where pressure was introduced, we suggested that the pressure in a gas is due to collisions of the molecules with the walls of its container. The force due to one such collision may be unmeasurably tiny, but the steady rain of a vast number of molecules striking a wall each second exerts a measurable macroscopic force. The gas pressure is the force per unit area ($p = F/A$) resulting from these molecular collisions.

Our task in this section is to calculate the pressure by doing the appropriate averaging over molecular motions and collisions. This task can be divided into three main pieces:

1. Calculate the impulse a single molecule exerts on the wall during a collision.
2. Find the force due to all collisions.
3. Introduce an appropriate average speed.

Force Due to a Single Collision

FIGURE 18.5 shows a molecule with an x-component of velocity v_x colliding with a wall and rebounding with its x-component of velocity changed from $+v_x$ to $-v_x$. This molecule experiences an impulse. We can use the impulse-momentum theorem from Chapter 9 to write

$$(J_x)_{\text{wall on molecule}} = \Delta p = m(-v_x) - mv_x = -2mv_x \qquad (18.4)$$

According to Newton's third law, the wall experiences the equal but opposite impulse

$$(J_x)_{\text{molecule on wall}} = +2mv_x \qquad (18.5)$$

as a result of this single collision.

Suppose there are N_{coll} such collisions during a very small time interval Δt. If we assume for the moment that all molecules have the *same* x-component velocity v_x, the net impulse of these collisions on the wall is

$$J_{\text{wall}} = N_{\text{coll}} \times (J_x)_{\text{molecule on wall}} = 2N_{\text{coll}}mv_x \qquad (18.6)$$

FIGURE 18.6 reminds you that impulse is the area under the force-versus-time curve and thus $J_{\text{wall}} = F_{\text{avg}}\Delta t$, where F_{avg} is the *average* force exerted on the wall. Using this in Equation 18.6, we see that the average force on the wall due to many molecular collisions is

$$F_{\text{avg}} = 2\frac{N_{\text{coll}}}{\Delta t}mv_x \qquad (18.7)$$

The quantity $N_{\text{coll}}/\Delta t$ is the *rate* of collisions with the wall—that is, the number of collisions per second. FIGURE 18.7 shows how to determine the rate of collisions. Let the time interval Δt be much less than the average time between molecular collisions, so no collisions alter the molecular speeds during this interval. (This assumption about Δt isn't really necessary, but it makes it easier to think about what's going on.) During Δt, all molecules travel distance $\Delta x = v_x\Delta t$ along the x-axis. This distance is shaded in the figure. *Every one* of the molecules in this shaded region that is moving to the right will reach and collide with the wall during time Δt. Molecules outside this region will not reach the wall during Δt and will not collide.

The shaded region has volume $A\Delta x$, where A is the surface area of the wall. Because of their random motions, only half the molecules are moving to the right, hence the number of collisions during Δt is

$$N_{\text{coll}} = \frac{1}{2}\frac{N}{V}A\Delta x = \frac{1}{2}\frac{N}{V}Av_x\Delta t \qquad (18.8)$$

and thus the rate of collisions is

$$\frac{N_{\text{coll}}}{\Delta t} = \frac{1}{2}\frac{N}{V}Av_x \qquad (18.9)$$

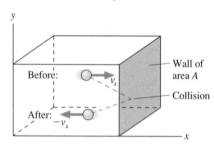

FIGURE 18.5 A molecule colliding with the wall exerts an impulse on it.

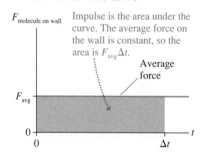

FIGURE 18.6 Impulse is the area under the force-versus-time curve.

Impulse is the area under the curve. The average force on the wall is constant, so the area is $F_{\text{avg}}\Delta t$.

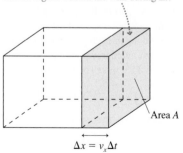

FIGURE 18.7 Determining the rate of collisions.

Only molecules moving to the right in the shaded region will hit the wall during Δt.

$\Delta x = v_x\Delta t$

The average force on the wall is found by substituting $N_{coll}/\Delta t$ from Equation 18.9 into Equation 18.7:

$$F_{avg} = 2\left(\frac{1}{2}\frac{N}{V}Av_x\right)mv_x = \frac{N}{V}mv_x^2 A \qquad (18.10)$$

Notice that this expression for F_{avg} does not depend on any details of the molecular collisions.

We can relax the assumption that all molecules have the same speed by replacing the squared velocity v_x^2 in Equation 18.10 with its average value. That is,

$$F_{avg} = \frac{N}{V}m(v_x^2)_{avg}A \qquad (18.11)$$

where $(v_x^2)_{avg}$ is the quantity v_x^2 averaged over all the molecules in the container.

The Root-Mean-Square Speed

We need to be somewhat careful when averaging velocities. The velocity component v_x has a sign. At any instant of time, half the molecules in a container move to the right and have positive v_x while the other half move to the left and have negative v_x. Thus the *average velocity* is $(v_x)_{avg} = 0$. If this weren't true, the entire container of gas would move away!

The speed of a molecule is $v = (v_x^2 + v_y^2 + v_z^2)^{1/2}$. Thus the average of the speed squared is

$$(v^2)_{avg} = (v_x^2 + v_y^2 + v_z^2)_{avg} = (v_x^2)_{avg} + (v_y^2)_{avg} + (v_z^2)_{avg} \qquad (18.12)$$

The square root of $(v^2)_{avg}$ is called the **root-mean-square speed** v_{rms}:

$$v_{rms} = \sqrt{(v^2)_{avg}} \qquad \text{(root-mean-square speed)} \qquad (18.13)$$

This is usually called the *rms speed*. You can remember its definition by noting that its name is the *opposite* of the sequence of operations: First you square all the speeds, then you average the squares (find the mean), then you take the square root. Because the square root "undoes" the square, v_{rms} must, in some sense, give an average speed.

> NOTE ▶ We could compute a true average speed v_{avg}, but that calculation is difficult. More important, the root-mean-square speed tends to arise naturally in many scientific and engineering calculations. It turns out that v_{rms} differs from v_{avg} by less than 10%, so for practical purposes we can interpret v_{rms} as being essentially the average speed of a molecule in a gas. ◀

EXAMPLE 18.2 **Calculating the root-mean-square speed**

FIGURE 18.8 shows the velocities of all the molecules in a six-molecule, two-dimensional gas. Calculate and compare the average velocity $\vec{v}_{avg}$, the average speed v_{avg}, and the rms speed v_{rms}.

SOLVE Table 18.1 on the next page shows the velocity components v_x and v_y for each molecule, the squares v_x^2 and v_y^2, their sum $v^2 = v_x^2 + v_y^2$, and the speed $v = (v_x^2 + v_y^2)^{1/2}$. Averages of all the values in each column are shown at the bottom. You can see that the average velocity is $\vec{v}_{avg} = \vec{0}$ m/s and the average speed is $v_{avg} = 11.9$ m/s. The rms speed is

$$v_{rms} = \sqrt{(v^2)_{avg}} = \sqrt{148.3 \text{ m}^2/\text{s}^2} = 12.2 \text{ m/s}$$

ASSESS The rms speed is only 2.5% greater than the average speed.

FIGURE 18.8 The molecular velocities of Example 18.2. Units are m/s.

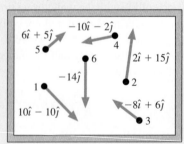

TABLE 18.1 Calculation of rms speed and average speed for the molecules of Example 18.2

Molecule	v_x	v_y	v_x^2	v_y^2	v^2	v
1	10	−10	100	100	200	14.1
2	2	15	4	225	229	15.1
3	−8	6	64	36	100	10.0
4	−10	−2	100	4	104	10.2
5	6	5	36	25	61	7.8
6	0	−14	0	196	196	14.0
Average	0	0			148.3	11.9

There's nothing special about the *x*-axis. The coordinate system is something that *we* impose on the problem, so *on average* it must be the case that

$$(v_x^2)_{\text{avg}} = (v_y^2)_{\text{avg}} = (v_z^2)_{\text{avg}} \tag{18.14}$$

Hence we can use Equation 18.12 and the definition of v_{rms} to write

$$v_{\text{rms}}^2 = (v_x^2)_{\text{avg}} + (v_y^2)_{\text{avg}} + (v_z^2)_{\text{avg}} = 3(v_x^2)_{\text{avg}} \tag{18.15}$$

Consequently, $(v_x^2)_{\text{avg}}$ is

$$(v_x^2)_{\text{avg}} = \frac{1}{3} v_{\text{rms}}^2 \tag{18.16}$$

Using this result in Equation 18.11 gives us the net force on the wall of the container:

$$F_{\text{net}} = \frac{1}{3} \frac{N}{V} m v_{\text{rms}}^2 A \tag{18.17}$$

Thus the pressure on the wall of the container due to all the molecular collisions is

$$p = \frac{F}{A} = \frac{1}{3} \frac{N}{V} m v_{\text{rms}}^2 \tag{18.18}$$

We have met our goal. Equation 18.18 expresses the macroscopic pressure in terms of the microscopic physics. The pressure depends on the number density of molecules in the container and on how fast, on average, the molecules are moving.

EXAMPLE 18.3 The rms speed of helium atoms

A container holds helium at a pressure of 200 kPa and a temperature of 60.0°C. What is the rms speed of the helium atoms?

SOLVE The rms speed can be found from the pressure and the number density. Using the ideal-gas law gives us the number density:

$$\frac{N}{V} = \frac{p}{k_B T} = \frac{200{,}000 \text{ Pa}}{(1.38 \times 10^{-23} \text{ J/K})(333 \text{ K})} = 4.35 \times 10^{25} \text{ m}^{-3}$$

The mass of a helium atom is $m = 4 \text{ u} = 6.64 \times 10^{-27}$ kg. Thus

$$v_{\text{rms}} = \sqrt{\frac{3p}{(N/V)m}} = 1440 \text{ m/s}$$

STOP TO THINK 18.2 The speed of every molecule in a gas is suddenly increased by a factor of 4. As a result, v_{rms} increases by a factor of

a. 2.
c. 4.
e. 16.

b. <4 but not necessarily 2.
d. >4 but not necessarily 16.
f. v_{rms} doesn't change.

18.3 Temperature

A molecule of mass m and velocity v has translational kinetic energy

$$\epsilon = \frac{1}{2}mv^2 \tag{18.19}$$

We'll use ϵ (lowercase Greek epsilon) to distinguish the energy of a molecule from the system energy E. Thus the average translational kinetic energy is

$$\epsilon_{avg} = \text{average translational kinetic energy of a molecule}$$

$$= \frac{1}{2}m(v^2)_{avg} = \frac{1}{2}mv_{rms}^2 \tag{18.20}$$

We've included the word "translational" to distinguish ϵ from rotational kinetic energy, which we will consider later in this chapter.

We can write the gas pressure, Equation 18.18, in terms of the average translational kinetic energy as

$$p = \frac{2}{3}\frac{N}{V}\left(\frac{1}{2}mv_{rms}^2\right) = \frac{2}{3}\frac{N}{V}\epsilon_{avg} \tag{18.21}$$

The pressure is directly proportional to the average molecular translational kinetic energy. This makes sense. More-energetic molecules will hit the walls harder as they bounce and thus exert more force on the walls.

It's instructive to write Equation 18.21 as

$$pV = \frac{2}{3}N\epsilon_{avg} \tag{18.22}$$

We know, from the ideal-gas law, that

$$pV = Nk_BT \tag{18.23}$$

Comparing these two equations, we reach the significant conclusion that the average translational kinetic energy per molecule is

$$\epsilon_{avg} = \frac{3}{2}k_BT \quad \text{(average translational kinetic energy)} \tag{18.24}$$

where the temperature T is in kelvins. For example, the average translational kinetic energy of a molecule at room temperature (20°C) is

$$\epsilon_{avg} = \frac{3}{2}(1.38 \times 10^{-23} \text{ J/K})(293 \text{ K}) = 6.1 \times 10^{-21} \text{ J}$$

NOTE ▶ A molecule's average translational kinetic energy depends *only* on the temperature, not on the molecule's mass. If two gases have the same temperature, their molecules have the same average translational kinetic energy. ◀

Equation 18.24 is especially satisfying because it finally gives real meaning to the concept of temperature. Writing it as

$$T = \frac{2}{3k_B}\epsilon_{avg} \tag{18.25}$$

we can see that, for a gas, **this thing we call *temperature* measures the average translational kinetic energy.** A higher temperature corresponds to a larger value of ϵ_{avg} and thus to higher molecular speeds. This concept of temperature also gives meaning to *absolute zero* as the temperature at which $\epsilon_{avg} = 0$ and all molecular motion ceases. (Quantum effects at very low temperatures prevent the motions from actually stopping, but our classical theory predicts that they would.) FIGURE 18.9 summarizes what we've learned thus far about the micro/macro connection.

FIGURE 18.9 The micro/macro connection for pressure and temperature.

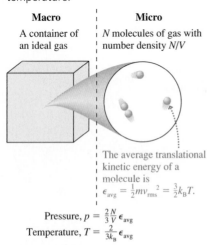

Macro	Micro
A container of an ideal gas	N molecules of gas with number density N/V

The average translational kinetic energy of a molecule is $\epsilon_{avg} = \frac{1}{2}mv_{rms}^2 = \frac{3}{2}k_BT$.

Pressure, $p = \frac{2}{3}\frac{N}{V}\epsilon_{avg}$

Temperature, $T = \frac{2}{3k_B}\epsilon_{avg}$

We can now justify our assumption that molecular collisions are perfectly elastic. Suppose they were not. If kinetic energy was lost in collisions, the average translational kinetic energy ϵ_{avg} of the gas would decrease and we would see a steadily decreasing temperature. But that doesn't happen. The temperature of an isolated system remains constant, indicating that ϵ_{avg} is not changing with time. Consequently, the collisions must be perfectly elastic.

EXAMPLE 18.4 | **Total microscopic kinetic energy**

What is the total translational kinetic energy of the molecules in 1.0 mol of gas at STP?

SOLVE The average translational kinetic energy of each molecule is

$$\epsilon_{avg} = \frac{3}{2}k_B T = \frac{3}{2}(1.38 \times 10^{-23} \text{ J/K})(273 \text{ K})$$

$$= 5.65 \times 10^{-21} \text{ J}$$

1.0 mol of gas contains N_A molecules; hence the total kinetic energy is

$$K_{micro} = N_A \epsilon_{avg} = 3400 \text{ J}$$

ASSESS The energy of any one molecule is incredibly small. Nonetheless, a macroscopic system has substantial thermal energy because it consists of an incredibly large number of molecules.

By definition, $\epsilon_{avg} = \frac{1}{2}mv_{rms}^2$. Using the ideal-gas law, we found $\epsilon_{avg} = \frac{3}{2}k_B T$. By equating these expressions we find that the rms speed of molecules in a gas is

$$v_{rms} = \sqrt{\frac{3k_B T}{m}} \qquad (18.26)$$

The rms speed depends on the square root of the temperature and inversely on the square root of the molecular mass.

EXAMPLE 18.5 | **Calculating an rms speed**

What is the rms speed of nitrogen molecules at room temperature (20°C)?

SOLVE The molecular mass is $m = 28 \text{ u} = 4.68 \times 10^{-26} \text{ kg}$ and $T = 20°C = 293 \text{ K}$. It is then a simple calculation to find

$$v_{rms} = \sqrt{\frac{3(1.38 \times 10^{-23} \text{ J/K})(293 \text{ K})}{4.68 \times 10^{-26} \text{ kg}}} = 509 \text{ m/s}$$

Some speeds will be greater than this and others smaller, but 509 m/s will be a typical or fairly average speed. This is in excellent agreement with the experimental results of Figure 18.2.

EXAMPLE 18.6 | **Mean time between collisions**

Estimate the mean time between collisions for a nitrogen molecule at 1.0 atm pressure and room temperature (20°C).

MODEL Because v_{rms} is essentially the average molecular speed, the *mean time between collisions* is simply the time needed to travel distance λ, the mean free path, at speed v_{rms}.

SOLVE We found $\lambda = 2.3 \times 10^{-7} \text{ m}$ in Example 18.1 and $v_{rms} = 509 \text{ m/s}$ in Example 18.5. Thus the mean time between collisions is

$$\tau_{coll} = \frac{\lambda}{v_{rms}} = \frac{2.3 \times 10^{-7} \text{ m}}{509 \text{ m/s}} = 4.5 \times 10^{-10} \text{ s}$$

ASSESS The air molecules around us move very fast, they collide with their neighbors about two billion times every second, and they manage to move, on average, only about 230 nm between collisions.

STOP TO THINK 18.3 The speed of every molecule in a gas is suddenly increased by a factor of 4. As a result, T increases by a factor of

a. 2.
b. <4 but not necessarily 2.
c. 4.
d. >4 but not necessarily 16.
e. 16.
f. T doesn't change.

18.4 Thermal Energy and Specific Heat

We defined the thermal energy of a system to be $E_{th} = K_{micro} + U_{micro}$, where K_{micro} is the microscopic kinetic energy of the moving molecules and U_{micro} is the potential energy of the stretched and compressed molecular bonds. We're now ready to take a microscopic look at thermal energy.

Monatomic Gases

FIGURE 18.10 shows a monatomic gas such as helium or neon. The atoms in an ideal gas have no molecular bonds with their neighbors; hence $U_{micro} = 0$. Furthermore, the kinetic energy of a monatomic gas particle is entirely translational kinetic energy ϵ. Thus the thermal energy of a monatomic gas of N atoms is

$$E_{th} = K_{micro} = \epsilon_1 + \epsilon_2 + \epsilon_3 + \cdots + \epsilon_N = N\epsilon_{avg} \tag{18.27}$$

where ϵ_i is the translational kinetic energy of atom i. We found that $\epsilon_{avg} = \frac{3}{2}k_B T$; hence the thermal energy is

$$E_{th} = \frac{3}{2}Nk_B T = \frac{3}{2}nRT \qquad \text{(thermal energy of a monatomic gas)} \tag{18.28}$$

where we used $N = nN_A$ and the definition of Boltzmann's constant, $k_B = R/N_A$.

We've noted for the last two chapters that thermal energy is associated with temperature. Now we have an explicit result for a monatomic gas: E_{th} is directly proportional to the temperature. Notice that E_{th} is independent of the atomic mass. Any two monatomic gases will have the same thermal energy if they have the same temperature and the same number of atoms (or moles).

If the temperature of a monatomic gas changes by ΔT, its thermal energy changes by

$$\Delta E_{th} = \frac{3}{2}nR\Delta T \tag{18.29}$$

In Chapter 17 we found that the change in thermal energy for *any* ideal-gas process is related to the molar specific heat at constant volume by

$$\Delta E_{th} = nC_V \Delta T \tag{18.30}$$

Equation 18.29 is a microscopic result that we obtained by relating the temperature to the average translational kinetic energy of the atoms. Equation 18.30 is a macroscopic result that we arrived at from the first law of thermodynamics. We can make a micro/macro connection by combining these two equations. Doing so gives us a *prediction* for the molar specific heat:

$$C_V = \frac{3}{2}R = 12.5 \text{ J/mol K} \qquad \text{(monatomic gas)} \tag{18.31}$$

This was exactly the value of C_V for all three monatomic gases in Table 17.4. The perfect agreement of theory and experiment is strong evidence that gases really do consist of moving, colliding molecules.

The Equipartition Theorem

The particles of a monatomic gas are atoms. Their energy consists exclusively of their translational kinetic energy. A particle's translational kinetic energy can be written

$$\epsilon = \frac{1}{2}mv^2 = \frac{1}{2}mv_x^2 + \frac{1}{2}mv_y^2 + \frac{1}{2}mv_z^2 = \epsilon_x + \epsilon_y + \epsilon_z \tag{18.32}$$

where we have written separately the energy associated with translational motion along the three axes. Because each axis in space is independent, we can think of ϵ_x, ϵ_y, and ϵ_z as independent *modes* of storing energy within the system.

FIGURE 18.10 The atoms in a monatomic gas have only translational kinetic energy.

Atom i has translational kinetic energy ϵ_i but no potential energy or rotational kinetic energy.

N atoms in a gas at temperature T

The thermal energy of the gas is $E_{th} = \epsilon_1 + \epsilon_2 + \epsilon_3 + \cdots = N\epsilon_{avg}$.

Other systems have additional modes of energy storage. For example,

- Two atoms joined by a spring-like molecular bond can vibrate back and forth. Both kinetic and potential energy are associated with this vibration.
- A diatomic molecule, in addition to translational kinetic energy, has rotational kinetic energy if it rotates end-over-end like a dumbbell.

We define the number of **degrees of freedom** as the number of distinct and independent modes of energy storage. A monatomic gas has three degrees of freedom, the three modes of translational kinetic energy. Systems that can vibrate or rotate have more degrees of freedom.

An important result of statistical physics says that the energy in a system is distributed so that all modes of energy storage have equal amounts of energy. This conclusion is known as the *equipartition theorem,* meaning that the energy is equally divided. The proof is beyond what we can do in this textbook, so we will state the theorem without proof:

Equipartition theorem The thermal energy of a system of particles is equally divided among all the possible degrees of freedom. For a system of N particles at temperature T, the energy stored in each mode (each degree of freedom) is $\frac{1}{2}Nk_BT$ or, in terms of moles, $\frac{1}{2}nRT$.

A monatomic gas has three degrees of freedom and thus, as we found above, $E_{th} = \frac{3}{2}Nk_BT$.

Solids

FIGURE 18.11 reminds you of our "bedspring model" of a solid with particle-like atoms connected by a lattice of spring-like molecular bonds. How many degrees of freedom does a solid have? Three degrees of freedom are associated with the kinetic energy, just as in a monatomic gas. In addition, the molecular bonds can be compressed or stretched independently along the x-, y-, and z-axes. Three additional degrees of freedom are associated with these three modes of potential energy. Altogether, a solid has six degrees of freedom.

The energy stored in each of these six degrees of freedom is $\frac{1}{2}Nk_BT$. The thermal energy of a solid is the total energy stored in all six modes, or

$$E_{th} = 3Nk_BT = 3nRT \qquad \text{(thermal energy of a solid)} \qquad (18.33)$$

We can use this result to predict the molar specific heat of a solid. If the temperature changes by ΔT, then the thermal energy changes by

$$\Delta E_{th} = 3nR\Delta T \qquad (18.34)$$

In Chapter 17 we defined the molar specific heat of a solid such that

$$\Delta E_{th} = nC\Delta T \qquad (18.35)$$

By comparing Equations 18.34 and 18.35 we can predict that the molar specific heat of a solid is

$$C = 3R = 25.0 \text{ J/mol K} \quad \text{(solid)} \qquad (18.36)$$

Not bad. The five elemental solids in Table 17.2 had molar specific heats clustered right around 25 J/mol K. They ranged from 24.3 J/mol K for aluminum to 26.5 J/mol K for lead. There are two reasons the agreement between theory and experiment isn't quite as perfect as it was for monatomic gases. First, our simple bedspring model of a solid isn't quite as accurate as our model of a monatomic gas. Second, quantum effects

FIGURE 18.11 A simple model of a solid.

Each atom has microscopic translational kinetic energy *and* microscopic potential energy along all three axes.

are beginning to make their appearance. More on this shortly. Nonetheless, our ability to predict C to within a few percent from a simple model of a solid is further evidence for the atomic structure of matter.

Diatomic Molecules

Diatomic molecules are a bigger challenge. How many degrees of freedom does a diatomic molecule have? FIGURE 18.12 shows a diatomic molecule, such as molecular nitrogen N_2, oriented along the x-axis. Three degrees of freedom are associated with the molecule's translational kinetic energy. The molecule can have a dumbbell-like end-over-end rotation about either the y-axis or the z-axis. It can also rotate about its own axis. These are three rotational degrees of freedom. The two atoms can also vibrate back and forth, stretching and compressing the molecular bond. This vibrational motion has both kinetic and potential energy—thus two more degrees of freedom.

Altogether, then, a diatomic molecule has eight degrees of freedom, and we would expect the thermal energy of a gas of diatomic molecules to be $E_{th} = 4k_BT$. The analysis we followed for a monatomic gas would then lead to the prediction $C_V = 4R = 33.2$ J/mol K. As compelling as this reasoning seems to be, this is *not* the experimental value of C_V that was reported for diatomic gases in Table 17.4. Instead, we found $C_V = 20.8$ J/mol K.

Why should a theory that works so well for monatomic gases and solids fail so miserably for diatomic molecules? To see what's going on, notice that 20.8 J/mol K = $\frac{5}{2}R$. A monatomic gas, with three degrees of freedom, has $C_V = \frac{3}{2}R$. A solid, with six degrees of freedom, has $C = 3R$. A diatomic gas would have $C_V = \frac{5}{2}R$ if it had five degrees of freedom, not eight.

This discrepancy was a major conundrum as statistical physics developed in the late 19th century. Although it was not recognized as such at the time, we are here seeing our first evidence for the breakdown of classical Newtonian physics. Classically, a diatomic molecule has eight degrees of freedom. The equipartition theorem doesn't distinguish between them; all eight should have the same energy. But atoms and molecules are not classical particles. It took the development of quantum theory in the 1920s to accurately characterize the behavior of atoms and molecules. We don't yet have the tools needed to see why, but quantum effects prevent three of the modes—the two vibrational modes and the rotation of the molecule about its own axis—from being active at room temperature.

FIGURE 18.13 shows C_V as a function of temperature for hydrogen gas. C_V is right at $\frac{5}{2}R$ for temperatures from $\approx$ 200 K up to $\approx$ 800 K. But at very low temperatures C_V drops to the monatomic-gas value $\frac{3}{2}R$. The two rotational modes become "frozen out" and the nonrotating molecule has only translational kinetic energy. Quantum physics can explain this, but not Newtonian physics. You can also see that the two vibrational modes *do* become active at very high temperatures, where C_V rises to

FIGURE 18.12 A diatomic molecule can rotate or vibrate.

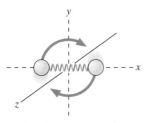

Rotation end-over-end about the z-axis

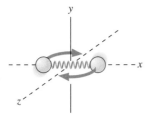

Rotation end-over-end about the y-axis

Rotation about the x-axis

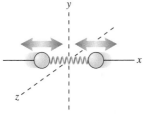

Vibration back and forth along the x-axis

FIGURE 18.13 Hydrogen molar specific heat at constant volume as a function of temperature. The temperature scale is logarithmic.

$\frac{7}{2}R$. Thus the real answer to What's wrong? is that Newtonian physics is not the right physics for describing atoms and molecules. We are somewhat fortunate that Newtonian physics is adequate to understand monatomic gases and solids, at least at room temperature.

Accepting the quantum result that a diatomic gas has only five degrees of freedom at commonly used temperatures (the translational degrees of freedom and the two end-over-end rotations), we find

$$E_{th} = \frac{5}{2}Nk_BT = \frac{5}{2}nRT$$

$$\text{(diatomic gases)} \qquad (18.37)$$

$$C_V = \frac{5}{2}R = 20.8 \text{ J/mol K}$$

A diatomic gas has more thermal energy than a monatomic gas at the same temperature because the molecules have rotational as well as translational kinetic energy.

While the micro/macro connection firmly establishes the atomic structure of matter, it also heralds the need for a new theory of matter at the atomic level. That is a task we will take up in Part VII. For now, Table 18.2 summarizes what we have learned from kinetic theory about thermal energy and molar specific heats.

TABLE 18.2 Kinetic theory predictions for the thermal energy and the molar specific heat

System	Degrees of freedom	E_{th}	C_V
Monatomic gas	3	$\frac{3}{2}Nk_BT = \frac{3}{2}nRT$	$\frac{3}{2}R = 12.5$ J/mol K
Diatomic gas	5	$\frac{5}{2}Nk_BT = \frac{5}{2}nRT$	$\frac{5}{2}R = 20.8$ J/mol K
Elemental solid	6	$3Nk_BT = 3nRT$	$3R = 25.0$ J/mol K

EXAMPLE 18.7 **The rotational frequency of a molecule**

The nitrogen molecule N_2 has a bond length of 0.12 nm. Estimate the rotational frequency of N_2 at 20°C.

MODEL The molecule can be modeled as a rigid dumbbell of length $L = 0.12$ nm rotating about its center.

SOLVE The rotational kinetic energy of the molecule is $\epsilon_{rot} = \frac{1}{2}I\omega^2$, where I is the moment of inertia about the center. Because we have two point masses each moving in a circle of radius $r = L/2$, the moment of inertia is

$$I = mr^2 + mr^2 = 2m\left(\frac{L}{2}\right)^2 = \frac{mL^2}{2}$$

Thus the rotational kinetic energy is

$$\epsilon_{rot} = \frac{1}{2}\frac{mL^2}{2}\omega^2 = \frac{mL^2\omega^2}{4} = \pi^2mL^2f^2$$

where we used $\omega = 2\pi f$ to relate the rotational frequency f to the angular frequency ω. From the equipartition theorem, the energy associated with this mode is $\frac{1}{2}Nk_BT$, so the *average* rotational kinetic energy per molecule is

$$(\epsilon_{rot})_{avg} = \frac{1}{2}k_BT$$

Equating these two expressions for ϵ_{rot} gives us

$$\pi^2mL^2f^2 = \frac{1}{2}k_BT$$

Thus the rotational frequency is

$$f = \sqrt{\frac{k_BT}{2\pi^2mL^2}} = 7.8 \times 10^{11} \text{ rev/s}$$

We evaluated f at $T = 293$ K, using $m = 14$ u $= 2.34 \times 10^{-26}$ kg for each *atom*.

ASSESS This is a *very* high frequency, but these values are typical of molecular rotations.

STOP TO THINK 18.4 How many degrees of freedom does a bead on a rigid rod have?

a. 1   b. 2   c. 3   d. 4   e. 5   f. 6

18.5 Thermal Interactions and Heat

FIGURE 18.14 Two gases can interact thermally through a very thin barrier.

Insulation prevents heat from entering or leaving the container.

A thin barrier prevents atoms from moving from system 1 to 2 but still allows them to collide. The barrier is clamped in place and cannot move.

FIGURE 18.14 Two gases can interact thermally through a very thin barrier.

We can now look in more detail at what happens when two systems at different temperatures interact with each other. FIGURE 18.14 shows a rigid, insulated container divided into two sections by a very thin, stiff membrane. The left side, which we'll call system 1, has N_1 atoms at an initial temperature T_{1i}. System 2 on the right has N_2 atoms at an initial temperature T_{2i}. The membrane is so thin that atoms can collide at the boundary as if the membrane were not there, yet it is a barrier that prevents atoms from moving from one side to the other. The situation is analogous, on an atomic scale, to basketballs colliding through a shower curtain.

Suppose that system 1 is initially at a higher temperature: $T_{1i} > T_{2i}$. This is not an equilibrium situation. The temperatures will change with time until the systems eventually reach a common final temperature T_f. If you *watch* the gases as one warms and the other cools, you see nothing happening. This interaction is quite different from a mechanical interaction in which, for example, you might see a piston move from one side toward the other. The only way in which the gases can interact is via molecular collisions at the boundary. This is a *thermal interaction,* and our goal is to understand how thermal interactions bring the systems to thermal equilibrium.

System 1 and system 2 begin with thermal energies

$$E_{1i} = \frac{3}{2}N_1 k_B T_{1i} = \frac{3}{2}n_1 R T_{1i}$$

$$E_{2i} = \frac{3}{2}N_2 k_B T_{2i} = \frac{3}{2}n_2 R T_{2i}$$

(18.38)

We've written the energies for monatomic gases; you could do the same calculation if one or both of the gases is diatomic by replacing the $\frac{3}{2}$ with $\frac{5}{2}$. Notice that we've omitted the subscript "th" to keep the notation manageable.

The total energy of the combined systems is $E_{tot} = E_{1i} + E_{2i}$. As systems 1 and 2 interact, their individual thermal energies E_1 and E_2 can change but their sum E_{tot} remains constant. The system will have reached thermal equilibrium when the individual thermal energies reach final values E_{1f} and E_{2f} that no longer change.

The Systems Exchange Energy

FIGURE 18.15 Collisions at the barrier transfer energy from faster molecules to slower molecules.

Thin barrier

Fast

Slow

Elastic collision

Loses energy

Gains energy

Energy transfer

FIGURE 18.15 shows a fast atom and a slow atom approaching the barrier from opposite sides. They undergo a perfectly elastic collision at the barrier. Although no net energy is lost in a perfectly elastic collision, the faster atom loses energy while the slower one gains energy. In other words, there is an energy *transfer* from the faster atom's side to the slower atom's side.

The average translational kinetic energy per molecule is directly proportional to the temperature: $\epsilon_{avg} = \frac{3}{2}k_B T$. Because $T_{1i} > T_{2i}$, the atoms in system 1 are, on average, more energetic than the atoms in system 2. Thus *on average* the collisions transfer energy from system 1 to system 2. Not in every collision: sometimes a fast atom in system 2 collides with a slow atom in system 1, transferring energy from 2 to 1. But the net energy transfer, from all collisions, is from the warmer system 1 to the cooler system 2. In other words, **heat is the energy transferred *via collisions* between the more-energetic (warmer) atoms on one side and the less-energetic (cooler) atoms on the other.**

How do the systems "know" when they've reached thermal equilibrium? Energy transfer continues until the atoms on both sides of the barrier have the *same average translational kinetic energy.* Once the average translational kinetic energies are the same, there is no tendency for energy to flow in either direction. This is the state of thermal equilibrium, so the condition for thermal equilibrium is

$$(\epsilon_1)_{avg} = (\epsilon_2)_{avg} \quad \text{(thermal equilibrium)} \quad (18.39)$$

where, as before, ϵ is the translational kinetic energy of an atom.

Because the average energies are directly proportional to the final temperatures, $\epsilon_{avg} = \frac{3}{2}k_B T_f$, thermal equilibrium is characterized by the macroscopic condition

$$T_{1f} = T_{2f} = T_f \qquad \text{(thermal equilibrium)} \qquad (18.40)$$

In other words, **two thermally interacting systems reach a common final temperature *because* they exchange energy via collisions until the atoms on each side have, on average, equal translational kinetic energies.** This is a very important idea.

Equation 18.40 can be used to determine the equilibrium thermal energies. Because these are monatomic gases, $E_{th} = N\epsilon_{avg}$. Thus the equilibrium condition $(\epsilon_1)_{avg} = (\epsilon_2)_{avg} = (\epsilon_{tot})_{avg}$ implies

$$\frac{E_{1f}}{N_1} = \frac{E_{2f}}{N_2} = \frac{E_{tot}}{N_1 + N_2} \qquad (18.41)$$

from which we can conclude

$$E_{1f} = \frac{N_1}{N_1 + N_2}E_{tot} = \frac{n_1}{n_1 + n_2}E_{tot}$$
$$\qquad\qquad\qquad\qquad\qquad\qquad (18.42)$$
$$E_{2f} = \frac{N_2}{N_1 + N_2}E_{tot} = \frac{n_2}{n_1 + n_2}E_{tot}$$

where in the last step we used moles rather than molecules.

Notice that $E_{1f} + E_{2f} = E_{tot}$, verifying that energy has been conserved even while being redistributed between the systems.

No work is done on either system because the barrier has no macroscopic displacement, so the first law of thermodynamics is

$$Q_1 = \Delta E_1 = E_{1f} - E_{1i}$$
$$\qquad\qquad\qquad\qquad\qquad (18.43)$$
$$Q_2 = \Delta E_2 = E_{2f} - E_{2i}$$

As a homework problem you can show that $Q_1 = -Q_2$, as required by energy conservation. That is, the heat lost by one system is gained by the other. $|Q_1|$ is the quantity of heat that is transferred from the warmer gas to the cooler gas during the thermal interaction.

NOTE ▶ In general, the equilibrium thermal energies of the system are *not* equal. That is, $E_{1f} \neq E_{2f}$. They will be equal only if $N_1 = N_2$. Equilibrium is reached when the average translational kinetic energies in the two systems are equal—that is, when $(\epsilon_1)_{avg} = (\epsilon_2)_{avg}$, not when $E_{1f} = E_{2f}$. The distinction is important. FIGURE 18.16 summarizes these ideas. ◀

FIGURE 18.16 Equilibrium is reached when the atoms on each side have, on average, equal energies.

Collisions transfer energy from the warmer system to the cooler system as more-energetic atoms lose energy to less-energetic atoms.

Thermal equilibrium occurs when the systems have the same average translational kinetic energy and thus the same temperature.

In general, the thermal energies E_{1f} and E_{2f} are *not* equal.

EXAMPLE 18.8 **A thermal interaction**

A sealed, insulated container has 2.0 g of helium at an initial temperature of 300 K on one side of a barrier and 10.0 g of argon at an initial temperature of 600 K on the other side.

a. How much heat energy is transferred, and in which direction?
b. What is the final temperature?

MODEL The systems start with different temperatures, so they are not in thermal equilibrium. Energy will be transferred via collisions from the argon to the helium until both systems have the same average molecular energy.

SOLVE a. Let the helium be system 1. Helium has molar mass $M_{mol} = 0.004$ kg/mol, so $n_1 = M/M_{mol} = 0.50$ mol. Similarly, argon has $M_{mol} = 0.040$ kg/mol, so $n_2 = 0.25$ mol. The initial thermal energies of the two monatomic gases are

$$E_{1i} = \frac{3}{2}n_1 R T_{1i} = 225R = 1870 \text{ J}$$

$$E_{2i} = \frac{3}{2}n_2 R T_{2i} = 225R = 1870 \text{ J}$$

The systems start with *equal* thermal energies, but they are not in thermal equilibrium. The total energy is $E_{tot} = 3740$ J.

Continued

In equilibrium, this energy is distributed between the two systems as

$$E_{1f} = \frac{n_1}{n_1 + n_2} E_{tot} = \frac{0.50}{0.75} 3740 \text{ J} = 2493 \text{ J}$$

$$E_{2f} = \frac{n_2}{n_1 + n_2} E_{tot} = \frac{0.25}{0.75} 3740 \text{ J} = 1247 \text{ J}$$

The heat entering or leaving each system is

$$Q_1 = Q_{He} = E_{1f} - E_{1i} = 623 \text{ J}$$

$$Q_2 = Q_{Ar} = E_{2f} - E_{2i} = -623 \text{ J}$$

The helium and the argon interact thermally via collisions at the boundary, causing 623 J of heat to be transferred from the warmer argon to the cooler helium.

b. These are constant-volume processes, thus $Q = nC_V\Delta T$. $C_V = \frac{3}{2}R$ for monatomic gases, so the temperature changes are

$$\Delta T_{He} = \frac{Q_{He}}{\frac{3}{2}nR} = \frac{623 \text{ J}}{1.5\,(0.50 \text{ mol})\,(8.31 \text{ J/mol K})} = 100 \text{ K}$$

$$\Delta T_{Ar} = \frac{Q_{Ar}}{\frac{3}{2}nR} = \frac{-623 \text{ J}}{1.5\,(0.25 \text{ mol})\,(8.31 \text{ J/mol K})} = -200 \text{ K}$$

Both gases reach the common final temperature $T_f = 400$ K.

ASSESS $E_{1f} = 2E_{2f}$ because there are twice as many atoms in system 1.

The main idea of this section is that two systems reach a common final temperature not by magic or by a prearranged agreement but simply from the energy exchange of vast numbers of molecular collisions. Real interacting systems, of course, are separated by walls rather than our unrealistic thin membrane. As the systems interact, the energy is first transferred via collisions from system 1 into the wall and subsequently, as the cooler molecules collide with a warm wall, into system 2. That is, the energy transfer is $E_1 \rightarrow E_{wall} \rightarrow E_2$. This is still heat because the energy transfer is occurring via molecular collisions rather than mechanical motion.

STOP TO THINK 18.5 Systems A and B are interacting thermally. At this instant of time,

a. $T_A > T_B$
b. $T_A = T_B$
c. $T_A < T_B$

A	B
$N = 1000$	$N = 2000$
$\epsilon_{avg} = 1.0 \times 10^{-20}$ J	$\epsilon_{avg} = 0.5 \times 10^{-20}$ J
$E_{th} = 1.0 \times 10^{-17}$ J	$E_{th} = 1.0 \times 10^{-17}$ J

18.6 Irreversible Processes and the Second Law of Thermodynamics

The preceding section looked at the thermal interaction between a warm gas and a cold gas. Heat energy is transferred from the warm gas to the cold gas until they reach a common final temperature. But why isn't heat transferred from the cold gas to the warm gas, making the cold side colder and the warm side warmer? Such a process could still conserve energy, but it never happens. The transfer of heat energy from hot to cold is an example of an **irreversible process,** a process that can happen only in one direction.

Examples of irreversible processes abound. Stirring the cream in your coffee mixes the cream and coffee together. No amount of stirring ever unmixes them. If you shake a jar that has red marbles on the top and blue marbles on the bottom, the two colors are quickly mixed together. No amount of shaking ever separates them again. If you watched a movie of someone shaking a jar and saw the red and blue marbles separating, you would be certain that the movie was running backward. In fact, a reasonable definition of an irreversible process is one for which a backward-running movie shows a physically impossible process.

FIGURE 18.17a is a two-frame movie of a collision between two particles, perhaps two gas molecules. Suppose that sometime after the collision is over we could reach in and reverse the velocities of both particles. That is, replace vector $\vec{v}$ with vector $-\vec{v}$. Then, as in a movie playing backward, the collision would happen in reverse. This is the movie of **FIGURE 18.17b**.

FIGURE 18.17 Molecular collisions are reversible.

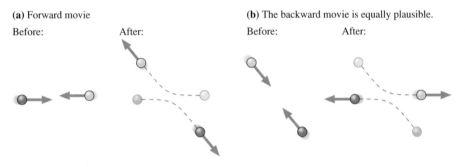

(a) Forward movie

Before: After:

(b) The backward movie is equally plausible.

Before: After:

You cannot tell, just by looking at the two movies, which is really going forward and which is being played backward. Maybe Figure 18.17b was the original collision and Figure 18.17a is the backward version. Nothing in either collision looks wrong, and no measurements you might make on either would reveal any violations of Newton's laws. Interactions at the molecular level are reversible processes.

Contrast this with the two-frame car crash movies in **FIGURE 18.18**. Past and future are clearly distinct in an irreversible process, and the backward movie of Figure 18.18b is obviously wrong. But what has been violated in the backward movie? To have the crumpled car spring away from the wall would not violate any laws of physics we have so far discovered. It would simply require transforming the thermal energy of the car and wall back into the macroscopic center-of-mass energy of the car as a whole.

The paradox stems from our assertion that macroscopic phenomena can be understood on the basis of microscopic molecular motions. If the microscopic motions are all reversible, how can the macroscopic phenomena end up being irreversible? If reversible collisions can cause heat to be transferred from hot to cold, why do they never cause heat to be transferred from cold to hot? There must be another law of physics preventing it. The law we seek must, in some sense, be able to distinguish the past from the future.

FIGURE 18.18 A car crash is irreversible.

(a) Forward movie

Before: After:

(b) The backward movie is physically impossible.

Before: After:

Which Way to Equilibrium?

Stated another way, how do two systems initially at different temperatures "know" which way to go to reach equilibrium? Perhaps an analogy will help.

FIGURE 18.19 shows two boxes, numbered 1 and 2, containing identical balls. Box 1 starts with more balls than box 2, so $N_{1i} > N_{2i}$. Once every second, one ball is chosen at random and moved to the other box. This is a reversible process because a ball can move from box 2 to box 1 just as easily as from box 1 to box 2. What do you expect to see if you return several hours later?

Because balls are chosen at random, and because $N_{1i} > N_{2i}$, it's initially more likely that a ball will move from box 1 to box 2 than from box 2 to box 1. Sometimes a ball will move "backward" from box 2 to box 1, but overall there's a net movement of balls from box 1 to box 2. The system will evolve until $N_1 \approx N_2$. This is a stable situation—equilibrium!—with an equal number of balls moving in both directions.

But couldn't it go the other way, with N_1 getting even larger while N_2 decreases? In principle, any possible arrangement of the balls is possible in the same way that any number of heads are possible if you throw N coins in the air and let them fall. If you

FIGURE 18.19 Two interacting systems. Balls are chosen at random and moved to the other box.

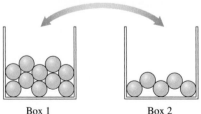

Balls are chosen at random and moved from one box to the other.

Box 1   Box 2
N_1 balls N_2 balls

throw four coins, the odds are 1 in 2^4, or 1 in 16, of getting four heads. With four balls, the odds are 1 in 16 that, at a randomly chosen instant of time, you would find $N_1 = 4$. You wouldn't find that to be terribly surprising.

With 10 balls, the probability that $N_1 = 10$ is $0.5^{10} \approx 1/1000$. With 100 balls, the probability that $N_1 = 100$ has dropped to $\approx 10^{-30}$. With 10^{20} balls, the odds of finding all of them, or even most of them, in one box are so staggeringly small that it's safe to say it will "never" happen. Although each transfer is reversible, **the statistics of large numbers make it overwhelmingly more likely that the system will evolve toward a state in which $N_1 \approx N_2$ than toward a state in which $N_1 > N_2$.**

The balls in our analogy represent energy. The total energy, like the total number of balls, is conserved, but molecular collisions can move energy between system 1 and system 2. Each collision is reversible, just as likely to transfer energy from 1 to 2 as from 2 to 1. But if $(\epsilon_{1i})_{avg} > (\epsilon_{2i})_{avg}$, and if we're dealing with two macroscopic systems where $N > 10^{20}$, then it's overwhelmingly likely that the net result of many, many collisions will be to transfer energy from system 1 to system 2 until $(\epsilon_{1f})_{avg} = (\epsilon_{2f})_{avg}$—in other words, for heat energy to be transferred from hot to cold.

The system reaches thermal equilibrium not by any plan or by outside intervention, but simply because **equilibrium is the *most probable* state in which to be.** It is *possible* that the system will move away from equilibrium, with heat moving from cold to hot, but remotely improbable in any realistic system. The consequence of a vast number of random events is that the system evolves in one direction, toward equilibrium, and not the other. **Reversible microscopic events lead to irreversible macroscopic behavior because some macroscopic states are vastly more probable than others.**

Order, Disorder, and Entropy

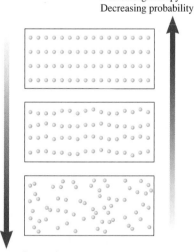

FIGURE 18.20 Ordered and disordered systems.

Increasing order
Decreasing entropy
Decreasing probability

Increasing randomness
Increasing entropy
Increasing probability

FIGURE 18.20 shows three different systems. At the top is a group of atoms arranged in a crystal-like lattice. This is a highly ordered and nonrandom system, with each atom's position precisely specified. Contrast this with the system on the bottom, where there is no order at all. The position of every atom was assigned entirely at random.

It is extremely improbable that the atoms in a container would *spontaneously* arrange themselves into the ordered pattern of the top picture. In a system of, say, 10^{20} atoms, the probability of this happening is similar to the probability that 10^{20} tossed coins will all be heads. We can safely say that it will never happen. By contrast, there are a vast number of arrangements like the one on the bottom that randomly fill the container.

The middle picture of Figure 18.20 is an in-between situation. This situation might arise as a solid melts. The positions of the atoms are clearly not completely random, so the system preserves some degree of order. This in-between situation is more likely to occur spontaneously than the highly ordered lattice on the top, but is less likely to occur than the completely random system on the bottom.

Scientists and engineers use a state variable called **entropy** to measure the probability that a macroscopic state will occur spontaneously. The ordered lattice, which has a very small probability of spontaneous occurrence, has a very low entropy. The entropy of the randomly filled container is high. The entropy of the middle picture is somewhere in between. It is often said that entropy measures the amount of *disorder* in a system. The entropy in Figure 18.20 increases as you move from the ordered system on the top to the disordered system on the bottom.

Similarly, two thermally interacting systems with different temperatures have a low entropy. These systems are ordered in the sense that the faster atoms are on one side of the barrier, the slower atoms on the other. The most random possible distribution of energy, and hence the least ordered system, corresponds to the situation where the two systems are in thermal equilibrium with equal temperatures. Entropy increases as two systems with initially different temperatures move toward equilibrium. Entropy would decrease if heat energy moved from cold to hot, making the hot system hotter and the cold system colder.

Entropy can be calculated, but we'll leave that to more advanced courses. For our purposes, the *concept* of entropy as a measure of the disorder in a system, or of the probability that a macroscopic state will occur, is more important than a numerical value.

The Second Law of Thermodynamics

The fact that macroscopic systems evolve irreversibly toward equilibrium is a statement about nature that is not contained in any of the laws of physics we have encountered. It is, in fact, a new law of physics, one known as the **second law of thermodynamics.**

The formal statement of the second law of thermodynamics is given in terms of entropy:

Tossing all heads, while not impossible, is extremely unlikely, and the probability of doing so rapidly decreases as the number of coins increases.

> **Second law, formal statement** The entropy of an isolated system (or group of systems) never decreases. The entropy either increases, until the system reaches equilibrium, or, if the system began in equilibrium, stays the same.

The qualifier "isolated" is most important. We can order the system by reaching in from the outside, perhaps using tiny tweezers to place the atoms in a lattice. Similarly, we can transfer heat from cold to hot by using a refrigerator. The second law is about what a system can or cannot do *spontaneously,* on its own, without outside intervention.

The second law of thermodynamics tells us that an isolated system evolves such that

- Order turns into disorder and randomness.
- Information is lost rather than gained.
- The system "runs down."

An isolated system never spontaneously generates order out of randomness. It is not that the system "knows" about order or randomness, but rather that there are vastly more states corresponding to randomness than there are corresponding to order. As collisions occur at the microscopic level, the laws of probability dictate that the system will, on average, move inexorably toward the most probable and thus most random macroscopic state.

The second law of thermodynamics is often stated in several equivalent but more informal versions. One of these, and the one most relevant to our discussion, is

> **Second law, informal statement #1** When two systems at different temperatures interact, heat energy is transferred spontaneously from the hotter to the colder system, never from the colder to the hotter.

The second law of thermodynamics is an independent statement about nature, separate from the first law. The first law is a precise statement about energy conservation. The second law, by contrast, is a *probabilistic* statement, based on the statistics of very large numbers. While it is conceivable that heat could spontaneously move from cold to hot, it will never occur in any realistic macroscopic system.

The irreversible evolution from less-likely macroscopic states to more-likely macroscopic states is what gives us a macroscopic direction of time. Stirring blends your coffee and cream, it never unmixes them. Friction causes an object to stop while increasing its thermal energy; the random atomic motions of thermal energy never spontaneously organize themselves into a macroscopic motion of the entire object.

A plant in a sealed jar dies and decomposes to carbon and various gases; the gases and carbon never spontaneously assemble themselves into a flower. These are all examples of irreversible processes. They each show a clear direction of time, a distinct difference between past and future.

Thus another statement of the second law is

> **Second law, informal statement #2** The time direction in which the entropy of an isolated macroscopic system increases is "the future."

Establishing the "arrow of time" is one of the most profound implications of the second law of thermodynamics.

The second law of thermodynamics has important implications for issues ranging from how we as a society use energy and resources to biological evolution and the future of the universe. We'll return to some of these issues in the Summary to Part IV. In the meantime, the second law will be used in Chapter 19 to understand some of the practical aspects of the thermodynamics of engines.

STOP TO THINK 18.6 Two identical boxes each contain 1,000,000 molecules. In box A, 750,000 molecules happen to be in the left half of the box while 250,000 are in the right half. In box B, 499,900 molecules happen to be in the left half of the box while 500,100 are in the right half. At this instant of time,

a. The entropy of box A is larger than the entropy of box B.
b. The entropy of box A is equal to the entropy of box B.
c. The entropy of box A is smaller than the entropy of box B.

SUMMARY

The goal of Chapter 18 has been to understand a macroscopic system in terms of the microscopic behavior of its molecules.

General Principles

The **micro/macro connection** relates the macroscopic properties of a system to the motion and collisions of its atoms and molecules.

The Equipartition Theorem

Tells us how collisions distribute the energy in the system. The energy stored in each mode of the system (each **degree of freedom**) is $\frac{1}{2}Nk_BT$ or, in terms of moles, $\frac{1}{2}nRT$.

The Second Law of Thermodynamics

Tells us how collisions move a system toward equilibrium. The entropy of an isolated system can only increase or, in equilibrium, stay the same.

- Order turns into disorder and randomness.

- Systems run down.

- Heat energy is transferred spontaneously from a hotter to a colder system, never from colder to hotter.

Important Concepts

Pressure is due to the force of the molecules colliding with the walls:

$$p = \frac{1}{3}\frac{N}{V}mv_{rms}^2 = \frac{2}{3}\frac{N}{V}\epsilon_{avg}$$

The thermal energy of a system is

E_{th} = translational kinetic energy + rotational kinetic energy + vibrational energy

- **Monatomic gas** $E_{th} = \frac{3}{2}Nk_BT = \frac{3}{2}nRT$
- **Diatomic gas** $E_{th} = \frac{5}{2}Nk_BT = \frac{5}{2}nRT$
- **Elemental solid** $E_{th} = 3Nk_BT = 3nRT$

The average translational kinetic energy of a molecule is

$\epsilon_{avg} = \frac{3}{2}k_BT$. The temperature of the gas $T = \frac{2}{3k_B}\epsilon_{avg}$

measures the average translational kinetic energy.

Heat is energy transferred via collisions from more-energetic molecules on one side to less-energetic molecules on the other. Equilibrium is reached when $(\epsilon_1)_{avg} = (\epsilon_2)_{avg}$, which implies $T_{1f} = T_{2f}$.

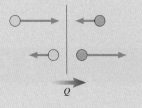

Entropy measures the probability that a macroscopic state will occur or, equivalently, the amount of disorder in a system.

Increasing entropy ➡

Applications

The root-mean-square speed v_{rms} is the square root of the average of the squares of the molecular speeds:

$$v_{rms} = \sqrt{(v^2)_{avg}}$$

For molecules of mass m at temperature T, $v_{rms} = \sqrt{\dfrac{3k_BT}{m}}$

Molar specific heats can be predicted from the thermal energy because $\Delta E_{th} = nC\Delta T$.

- **Monatomic gas** $C_V = \frac{3}{2}R$
- **Diatomic gas** $C_V = \frac{5}{2}R$
- **Elemental solid** $C = 3R$

Terms and Notation

histogram
mean free path, λ
root-mean-square speed, v_{rms}

degrees of freedom
equipartition theorem
irreversible process

entropy
second law of thermodynamics

CONCEPTUAL QUESTIONS

1. Solids and liquids resist being compressed. They are not totally incompressible, but it takes large forces to compress them even slightly. If it is true that matter consists of atoms, what can you infer about the microscopic nature of solids and liquids from their incompressibility?

2. Gases, in contrast with solids and liquids, are very compressible. What can you infer from this observation about the microscopic nature of gases?

3. The density of air at STP is about $\frac{1}{1000}$ the density of water. How does the average distance between air molecules compare to the average distance between water molecules? Explain.

4. The mean free path of molecules in a gas is 200 nm.
   a. What will be the mean free path if the pressure is doubled while all other state variables are held constant?
   b. What will be the mean free path if the absolute temperature is doubled while all other state variables are held constant?

5. If the pressure of a gas is really due to the *random* collisions of molecules with the walls of the container, why do pressure gauges—even very sensitive ones—give perfectly steady readings? Shouldn't the gauge be continually jiggling and fluctuating? Explain.

6. Suppose you could suddenly increase the speed of every molecule in a gas by a factor of 2.
   a. Would the rms speed of the molecules increase by a factor of $2^{1/2}$, 2, or 2^2? Explain.
   b. Would the gas pressure increase by a factor of $2^{1/2}$, 2, or 2^2? Explain.

7. Suppose you could suddenly increase the speed of every molecule in a gas by a factor of 2.
   a. Would the temperature of the gas increase by a factor of $2^{1/2}$, 2, or 2^2? Explain.

   b. Would the molar specific heat at constant volume change? If so, by what factor? If not, why not?

8. The two containers of gas in FIGURE Q18.8 are in good thermal contact with each other but well insulated from the environment. They have been in contact for a long time and are in thermal equilibrium.
   a. Is v_{rms} of helium greater than, less than, or equal to v_{rms} of argon? Explain.
   b. Does the helium have more thermal energy, less thermal energy, or the same amount of thermal energy as the argon? Explain.

FIGURE Q18.8

9. Suppose you place an ice cube in a beaker of room-temperature water, then seal them in a rigid, well-insulated container. No energy can enter or leave the container.
   a. If you open the container an hour later, will you find a beaker of water slightly cooler than room temperature, or a large ice cube and some 100°C steam?
   b. Finding a large ice cube and some 100°C steam would not violate the first law of thermodynamics. $W = 0$ J and $Q = 0$ J because the container is sealed, and $\Delta E_{th} = 0$ J because the increase in thermal energy of the water molecules that became steam is offset by the decrease in thermal energy of the water molecules that turned to ice. Energy would be conserved, yet we never see an outcome like this. Why not?

EXERCISES AND PROBLEMS

Problems labeled [   ] integrate material from earlier chapters.

Exercises

Section 18.1 Molecular Speeds and Collisions

1. | A 1.0 m × 1.0 m × 1.0 m cube of nitrogen gas is at 20°C and 1.0 atm. Estimate the number of molecules in the cube with a speed between 700 m/s and 1000 m/s.

2. | The number density of an ideal gas at STP is called the *Loschmidt number*. Calculate the Loschmidt number.

3. ‖ At what pressure will the mean free path in room-temperature (20°C) nitrogen be 1.0 m?

4. ‖ Integrated circuits are manufactured in vacuum chambers in which the air pressure is 1.0×10^{-10} mm of Hg. What are (a) the number density and (b) the mean free path of a molecule? Assume $T = 20°C$.

5. | The mean free path of a molecule in a gas is 300 nm. What will the mean free path be if the gas temperature is doubled at (a) constant volume and (b) constant pressure?

6. ‖ For a monatomic gas, what is the ratio of the volume per atom (V/N) to the volume *of* an atom when the mean free path is ten times the atomic diameter?

7. ‖ A lottery machine uses blowing air to keep 2000 Ping-Pong balls bouncing around inside a 1.0 m × 1.0 m × 1.0 m box. The diameter of a Ping-Pong ball is 3.0 cm. What is the mean free path between collisions? Give your answer in cm.

Section 18.2 Pressure in a Gas

8. | Eleven molecules have speeds 15, 16, 17, ... , 25 m/s. Calculate (a) v_{avg} and (b) v_{rms}.

9. ‖ The molecules in a six-particle gas have velocities

$$\vec{v}_1 = (20\hat{\imath} - 30\hat{\jmath}) \text{ m/s} \qquad \vec{v}_4 = 30\hat{\imath} \text{ m/s}$$

$$\vec{v}_2 = (40\hat{\imath} + 70\hat{\jmath}) \text{ m/s} \qquad \vec{v}_5 = (40\hat{\imath} - 40\hat{\jmath}) \text{ m/s}$$

$$\vec{v}_3 = (-80\hat{\imath} + 20\hat{\jmath}) \text{ m/s} \qquad \vec{v}_6 = (-50\hat{\imath} - 20\hat{\jmath}) \text{ m/s}$$

Calculate (a) $\vec{v}_{avg}$, (b) v_{avg}, and (c) v_{rms}.

10. | FIGURE EX18.10 is a histogram showing the speeds of the molecules in a very small gas. What are (a) the most probable speed, (b) the average speed, and (c) the rms speed?

FIGURE EX18.10

11. ‖ The number density in a container of neon gas is 5.00×10^{25} m^{-3}. The atoms are moving with an rms speed of 660 m/s. What are (a) the temperature and (b) the pressure inside the container?

12. ‖ A cylinder contains gas at a pressure of 2.0 atm and a number density of 4.2×10^{25} m^{-3}. The rms speed of the atoms is 660 m/s. Identify the gas.

13. ‖ At 100°C the rms speed of nitrogen molecules is 576 m/s. Nitrogen at 100°C and a pressure of 2.0 atm is held in a container with a 10 cm $\times$ 10 cm square wall. Estimate the rate of molecular collisions (collisions/s) on this wall.

Section 18.3 Temperature

14. ‖ What are the rms speeds of (a) argon atoms and (b) hydrogen molecules at 800°C?

15. ‖ A gas consists of a mixture of neon and argon. The rms speed of the neon atoms is 400 m/s. What is the rms speed of the argon atoms?

16. | 1.5 m/s is a typical walking speed. At what temperature (in °C) would nitrogen molecules have an rms speed of 1.5 m/s?

17. ‖ At what temperature (in °C) do hydrogen molecules have the same rms speed as nitrogen molecules at 100°C?

18. | At what temperature (in °C) is the rms speed of helium atoms (a) half and (b) twice its value at STP?

19. | The rms speed of molecules in a gas is 600 m/s. What will be the rms speed if the gas pressure and volume are both halved?

20. ‖ By what factor does the rms speed of a molecule change if the temperature is increased from 10°C to 1000°C?

21. ‖ Atoms can be "cooled" to incredibly low temperatures by letting them interact with a laser beam. Various novel quantum phenomena appear at these temperatures. What is the rms speed of cesium atoms that have been cooled to a temperature of 100 nK?

22. | At STP, what is the total translational kinetic energy of the molecules in 1.0 mol of (a) hydrogen, (b) helium, and (c) oxygen?

23. | Suppose you double the temperature of a gas at constant volume. Do the following change? If so, by what factor?
   a. The average translational kinetic energy of a molecule.
   b. The rms speed of a molecule.
   c. The mean free path.

24. ‖ During a physics experiment, helium gas is cooled to a temperature of 10 K at a pressure of 0.10 atm. What are (a) the mean free path in the gas, (b) the rms speed of the atoms, and (c) the average energy per atom?

25. | What are (a) the average kinetic energy and (b) the rms speed of a proton in the center of the sun, where the temperature is 2.0×10^7 K?

26. | The atmosphere of the sun consists mostly of hydrogen *atoms* (not molecules) at a temperature of 6000 K. What are (a) the average translational kinetic energy per atom and (b) the rms speed of the atoms?

Section 18.4 Thermal Energy and Specific Heat

27. ‖ A 10 g sample of neon gas has 1700 J of thermal energy. Estimate the average speed of a neon atom.

28. ‖ The rms speed of the atoms in a 2.0 g sample of helium gas is 700 m/s. What is the thermal energy of the gas?

29. ‖ A 6.0 m $\times$ 8.0 m $\times$ 3.0 m room contains air at 20°C. What is the room's thermal energy?

30. | The thermal energy of 1.0 mol of a substance is increased by 1.0 J. What is the temperature change if the system is (a) a monatomic gas, (b) a diatomic gas, and (c) a solid?

31. ‖ What is the thermal energy of 100 cm^3 of aluminum at 100°C?

32. | 1.0 mol of a monatomic gas interacts thermally with 1.0 mol of an elemental solid. The gas temperature decreases by 50°C at constant volume. What is the temperature change of the solid?

33. ‖ A cylinder of nitrogen gas has a volume of 15,000 cm^3 and a pressure of 100 atm.
   a. What is the thermal energy of this gas at room temperature (20°C)?
   b. What is the mean free path in the gas?
   c. The valve is opened and the gas is allowed to expand slowly and isothermally until it reaches a pressure of 1.0 atm. What is the change in the thermal energy of the gas?

34. ‖ A rigid container holds 0.20 g of hydrogen gas. How much heat is needed to change the temperature of the gas
   a. From 50 K to 100 K?
   b. From 250 K to 300 K?
   c. From 2250 K to 2300 K?

Section 18.5 Thermal Interactions and Heat

35. ‖ 4.0 mol of monatomic gas A interacts with 3.0 mol of monatomic gas B. Gas A initially has 9000 J of thermal energy, but in the process of coming to thermal equilibrium it transfers 1000 J of heat energy to gas B. How much thermal energy did gas B have initially?

36. | 2.0 mol of monatomic gas A initially has 5000 J of thermal energy. It interacts with 3.0 mol of monatomic gas B, which initially has 8000 J of thermal energy.
   a. Which gas has the higher initial temperature?
   b. What is the final thermal energy of each gas?

Problems

37. ‖‖ The pressure inside a tank of neon is 150 atm. The temperature is 25°C. On average, how many atomic diameters does a neon atom move between collisions?

38. ‖ From what height must an oxygen molecule fall in a vacuum so that its kinetic energy at the bottom equals the average energy of an oxygen molecule at 300 K?

39. ‖ A gas at $p = 50$ kPa and $T = 300$ K has a mass density of 0.0802 kg/m^3.
   a. Identify the gas.
   b. What is the rms speed of the atoms in this gas?
   c. What is the mean free path of the atoms in the gas?

40. ‖ Dust particles are $\approx 10\ \mu m$ in diameter. They are pulverized rock, with $\rho \approx 2500\ kg/m^3$. If you treat dust as an ideal gas, what is the rms speed of a dust particle at 20°C?

41. ‖ Interstellar space, far from any stars, is filled with a very low density of hydrogen atoms (H, not H_2). The number density is about 1 atom/cm³ and the temperature is about 3 K.
   a. Estimate the pressure in interstellar space. Give your answer in Pa and in atm.
   b. What is the rms speed of the atoms?
   c. What is the edge length L of an $L \times L \times L$ cube of gas with 1.0 J of thermal energy?

42. ‖ Equation 18.3 is the mean free path of a particle through a gas of identical particles of equal radius. An electron can be thought of as a point particle with zero radius.
   a. Find an expression for the mean free path of an electron through a gas.
   b. Electrons travel 3 km through the Stanford Linear Accelerator. In order for scattering losses to be negligible, the pressure inside the accelerator tube must be reduced to the point where the mean free path is at least 50 km. What is the maximum possible pressure inside the accelerator tube, assuming $T = 20°C$? Give your answer in both Pa and atm.

43. ‖ Uranium has two naturally occurring isotopes. ^{238}U has a natural abundance of 99.3% and ^{235}U has an abundance of 0.7%. It is the rarer ^{235}U that is needed for nuclear reactors. The isotopes are separated by forming uranium hexafluoride, UF_6, which is a gas, then allowing it to diffuse through a series of porous membranes. $^{235}UF_6$ has a slightly larger rms speed than $^{238}UF_6$ and diffuses slightly faster. Many repetitions of this procedure gradually separate the two isotopes. What is the ratio of the rms speed of $^{235}UF_6$ to that of $^{238}UF_6$?

44. ‖ On earth, STP is based on the average atmospheric pressure at the surface and on a phase change of water that occurs at an easily produced temperature, being only slightly cooler than the average air temperature. The atmosphere of Venus is almost entirely carbon dioxide (CO_2), the pressure at the surface is a staggering 93 atm, and the average temperature is 470°C. Venusian scientists, if they existed, would certainly use the surface pressure as part of their definition of STP. To complete the definition, they would seek a phase change that occurs near the average temperature. Conveniently, the melting point of the element tellurium is 450°C. What are (a) the rms speed and (b) the mean free path of carbon dioxide molecules at Venusian STP based on this phase change in tellurium. The radius of a CO_2 molecule is 1.5×10^{-10} m.

45. ‖‖ 5.0×10^{23} nitrogen molecules collide with a 10 cm² wall each second. Assume that the molecules all travel with a speed of 400 m/s and strike the wall head-on. What is the pressure on the wall?

46. ‖‖‖ A 10 cm × 10 cm × 10 cm box contains 0.010 mol of nitrogen at 20°C. What is the rate of collisions (collisions/s) on one wall of the box?

47. ‖ FIGURE P18.47 shows the thermal energy of 0.14 mol of gas as a function of temperature. What is C_V for this gas?

FIGURE P18.47

48. ‖ A 100 cm³ box contains helium at a pressure of 2.0 atm and a temperature of 100°C. It is placed in thermal contact with a 200 cm³ box containing argon at a pressure of 4.0 atm and a temperature of 400°C.
   a. What is the initial thermal energy of each gas?
   b. What is the final thermal energy of each gas?
   c. How much heat energy is transferred, and in which direction?
   d. What is the final temperature?
   e. What is the final pressure in each box?

49. ‖ 2.0 g of helium at an initial temperature of 300 K interacts thermally with 8.0 g of oxygen at an initial temperature of 600 K.
   a. What is the initial thermal energy of each gas?
   b. What is the final thermal energy of each gas?
   c. How much heat energy is transferred, and in which direction?
   d. What is the final temperature?

50. ‖ A gas of 1.0×10^{20} atoms or molecules has 1.0 J of thermal energy. Its molar specific heat at constant pressure is 20.8 J/mol K. What is the temperature of the gas?

51. ‖ How many degrees of freedom does a system have if $\gamma = 1.29$?

52. ‖ A monatomic gas and a diatomic gas have equal numbers of moles and equal temperatures. Both are heated at constant pressure until their volume doubles. What is the ratio $Q_{diatomic}/Q_{monatomic}$?

53. ‖ In the discussion following Equation 18.43 it was said that $Q_1 = -Q_2$. Prove that this is so.

54. ‖ A monatomic gas is adiabatically compressed to $\frac{1}{8}$ of its initial volume. Does each of the following quantities change? If so, does it increase or decrease, and by what factor? If not, why not?
   a. The rms speed.
   b. The mean free path.
   c. The thermal energy of the gas.
   d. The molar specific heat at constant volume.

55. ‖ A diatomic gas is isobarically expanded to four times its initial volume. Does each of the following quantities change? If so, does it increase or decrease, and by what factor? If not, why not?
   a. The rms speed.
   b. The mean free path.
   c. The thermal energy of the gas.
   d. The molar heat capacity at constant volume.

56. ‖ The 2010 Nobel Prize in Physics was awarded for the discovery of graphene, a two-dimensional form of carbon in which the atoms form a two-dimensional crystal-lattice sheet only one atom thick. Predict the molar specific heat of graphene. Give your answer as a multiple of R.

57. ‖ Equal masses of hydrogen gas and oxygen gas are mixed together in a container and held at constant temperature. What is the hydrogen/oxygen ratio of (a) v_{rms}, (b) ϵ_{avg}, and (c) E_{th}?

58. ‖ The rms speed of the molecules in 1.0 g of hydrogen gas is 1800 m/s.
   a. What is the total translational kinetic energy of the gas molecules?
   b. What is the thermal energy of the gas?
   c. 500 J of work are done to compress the gas while, in the same process, 1200 J of heat energy are transferred from the gas to the environment. Afterward, what is the rms speed of the molecules?

59. ‖ At what temperature does the rms speed of (a) a nitrogen molecule and (b) a hydrogen molecule equal the escape speed from the earth's surface? (c) You'll find that these temperatures are very high, so you might think that the earth's gravity could easily contain both gases. But not all molecules move with v_{rms}. There

is a distribution of speeds, and a small percentage of molecules have speeds several times v_{rms}. Bit by bit, a gas can slowly leak out of the atmosphere as its fastest molecules escape. A reasonable rule of thumb is that the earth's gravity can contain a gas only if the average translational kinetic energy per molecule is less than 1% of the kinetic energy needed to escape. Use this rule to show why the earth's atmosphere contains nitrogen but not hydrogen, even though hydrogen is the most abundant element in the universe.

60. ‖ n_1 moles of a monatomic gas and n_2 moles of a diatomic gas are mixed together in a container.
   a. Derive an expression for the molar specific heat at constant volume of the mixture.
   b. Show that your expression has the expected behavior if $n_1 \rightarrow 0$ or $n_2 \rightarrow 0$.

61. ‖ A 1.0 kg ball is at rest on the floor in a 2.0 m × 2.0 m × 2.0 m room of air at STP. Air is 80% nitrogen (N_2) and 20% oxygen (O_2) by volume.
   a. What is the thermal energy of the air in the room?
   b. What fraction of the thermal energy would have to be conveyed to the ball for it to be spontaneously launched to a height of 1.0 m?
   c. By how much would the air temperature have to decrease to launch the ball?
   d. Your answer to part c is so small as to be unnoticeable, yet this event never happens. Why not?

62. ‖ An inventor wants you to invest money with his company, offering you 10% of all future profits. He reminds you that the brakes on cars get extremely hot when they stop and that there is a large quantity of thermal energy in the brakes. He has invented a device, he tells you, that converts that thermal energy into the forward motion of the car. This device will take over from the engine after a stop and accelerate the car back up to its original speed, thereby saving a tremendous amount of gasoline. Now, you're a smart person, so he admits up front that this device is not 100% efficient, that there is some unavoidable heat loss to the air and to friction within the device, but the upcoming research for which he needs your investment will make those losses extremely small. You do also have to start the car with cold brakes after it has been parked awhile, so you'll still need a

gasoline engine for that. Nonetheless, he tells you, his prototype car gets 500 miles to the gallon and he expects to be at well over 1000 miles to the gallon after the next phase of research. Should you invest? Base your answer on an analysis of the *physics* of the situation.

Challenge Problems

63. n moles of a diatomic gas with $C_V = \frac{5}{2}R$ has initial pressure p_i and volume V_i. The gas undergoes a process in which the pressure is directly proportional to the volume until the rms speed of the molecules has doubled.
   a. Show this process on a pV diagram.
   b. How much heat does this process require? Give your answer in terms of n, p_i, and V_i.

64. An experiment you're designing needs a gas with $\gamma = 1.50$. You recall from your physics class that no individual gas has this value, but it occurs to you that you could produce a gas with $\gamma = 1.50$ by mixing together a monatomic gas and a diatomic gas. What fraction of the molecules need to be monatomic?

65. Consider a container like that shown in Figure 18.14, with n_1 moles of a monatomic gas on one side and n_2 moles of a diatomic gas on the other. The monatomic gas has initial temperature T_{1i}. The diatomic gas has initial temperature T_{2i}.
   a. Show that the equilibrium thermal energies are

$$E_{1f} = \frac{3n_1}{3n_1 + 5n_2}(E_{1i} + E_{2i})$$

$$E_{2f} = \frac{5n_2}{3n_1 + 5n_2}(E_{1i} + E_{2i})$$

   b. Show that the equilibrium temperature is

$$T_f = \frac{3n_1 T_{1i} + 5n_2 T_{2i}}{3n_1 + 5n_2}$$

   c. 2.0 g of helium at an initial temperature of 300 K interacts thermally with 8.0 g of oxygen at an initial temperature of 600 K. What is the final temperature? How much heat energy is transferred, and in which direction?

STOP TO THINK ANSWERS

Stop to Think 18.1: $\lambda_B > \lambda_A = \lambda_C > \lambda_D$. Increasing the volume makes the gas less dense, so λ increases. Increasing the radius makes the targets larger, so λ decreases. The mean free path doesn't depend on the atomic mass.

Stop to Think 18.2: c. Each v^2 increases by a factor of 16 but, after averaging, v_{rms} takes the square root.

Stop to Think 18.3: e. Temperature is proportional to the average energy. The energy of a gas molecule is kinetic, proportional to v^2. The average energy, and thus T, increases by 4^2.

Stop to Think 18.4: b. The bead can slide along the wire (one degree of translational motion) and rotate around the wire (one degree of rotational motion).

Stop to Think 18.5: a. Temperature measures the average translational kinetic energy *per molecule,* not the thermal energy of the entire system.

Stop to Think 18.6: c. With 1,000,000 molecules, it's highly unlikely that 750,000 of them would spontaneously move into one side of the box. A state with a very small probability of occurrence has a very low entropy. Having an imbalance of only 100 out of 1,000,000 is well within what you might expect for random fluctuations. This is a highly probable situation and thus one of large entropy.

19 Heat Engines and Refrigerators

This power plant is generating electricity by turning heat into work—but not very efficiently. The cooling towers dissipate roughly two-thirds of the fuel's energy into the air as "waste heat."

▶ **Looking Ahead** The goal of Chapter 19 is to study the physical principles that govern heat engines and refrigerators.

Turning Heat into Work

Modern society is powered by devices that transform the heat energy of burning fuel into useful work, such as

- Pumping water.
- Propelling cars and airplanes.
- Generating electricity.

Our goal in this chapter is not to study specific devices but to look at the underlying physics that governs all such devices

You will learn that heat engines

- Follow a cyclical process that can be shown on pV diagrams and on *energy-transfer diagrams*.
- Require not only a source of heat but also a source of cooling. These are called the *hot reservoir* and the *cold reservoir*.
- Are governed by the first and second laws of thermodynamics.

◀ **Looking Back**
Sections 17.2–17.4 Work, heat, and the first law of thermodynamics

Heat Engines

Heat engine is the generic name for any device that uses a cyclical process to transform heat energy into work.

The heat from burning fuel boils water to make high-pressure steam that then does work by spinning this turbine at an electric generating station.

Efficiency

How good is a heat engine at transforming heat into work? We'll define an engine's **thermal efficiency** as

$$\text{efficiency} = \frac{\text{work done}}{\text{heat required}}$$

You'll learn that the laws of thermodynamics set limits on the maximum possible efficiency. The fact that no heat engine can have an efficiency of 100% prevents us from extracting and using the vast thermal energy in the air and water around us.

Refrigerators

A **refrigerator** is a heat engine in reverse, using work to "pump energy uphill" from cold to hot.

In a refrigerator, a compressor does work to pump heat energy from the colder inside to the warmer room. Air conditioners are "refrigerators" pumping heat energy from the cool inside of a house to the hot outside.

The Carnot Engine

We'll use the second law of thermodynamics to show that a *perfectly reversible heat engine*—called a **Carnot engine**—has the maximum possible thermal efficiency.

You'll learn that the efficiency of a Carnot engine depends only on the temperatures of the hot and cold reservoirs. Any real engine's efficiency will be less—often much less.

◀ **Looking Back**
Section 18.6 The second law of thermodynamics

19.1 Turning Heat into Work

Thermodynamics is the branch of physics that studies the transformation of energy. Many practical devices are designed to transform energy from one form, such as the heat from burning fuel, into another, such as work. Chapters 17 and 18 established two laws of thermodynamics that any such device must obey:

> **First law** Energy is conserved; that is, $\Delta E_{th} = W + Q$.
>
> **Second law** Most macroscopic processes are irreversible. In particular, heat energy is transferred spontaneously from a hotter to a colder system but never from a colder system to a hotter system.

Our goal in this chapter is to discover what these two laws, especially the second law, imply about devices that turn heat into work. In particular:

■ How does a practical device transform heat into work?
■ What are the limitations and restrictions on these energy transformations?

Work Done by the System

The work W in the first law is the work done *on* the system by external forces from the environment. However, it makes more sense in "practical thermodynamics" to use the work done *by* the system. For example, you want to know how much useful work you can obtain from an expanding gas. The work done by the system is called W_s.

Work done by the environment and work done by the system are not mutually exclusive. In fact, they are very simply related by $W_s = -W$. In FIGURE 19.1, force $\vec{F}_{gas}$ due to the gas pressure does work when the piston moves. This is W_s, the work done *by* the system. At the same time, some object in the environment, such as a piston rod, must be pushing inward with force $\vec{F}_{ext} = -\vec{F}_{gas}$ to keep the gas pressure from blowing the piston out. This force does the work W *on* the system, work that you've learned is the negative of the area under the pV curve of the process.

Because the forces are equal but opposite, we see that

$$W_s = -W = \text{the area under the } pV \text{ curve} \tag{19.1}$$

When a gas expands and pushes the piston out, transferring energy out of the system, we say "the system does work on the environment." While this may seem to imply that the environment is doing no work on the system, all the phrase means is that W_s is positive and W is negative.

Similarly, "the environment does work on the system" means that $W > 0$ (energy is transferred into the system) and thus $W_s < 0$. Whether we use W or W_s is a matter of convenience. They are always opposite to each other rather than one being zero.

The first law of thermodynamics $\Delta E_{th} = W + Q$ can be written in terms of W_s as

$$Q = W_s + \Delta E_{th} \quad \text{(first law of thermodynamics)} \tag{19.2}$$

Any energy transferred into a system as heat is either used to do work or stored within the system as an increased thermal energy.

Energy-Transfer Diagrams

Suppose you drop a hot rock into the ocean. Heat is transferred from the rock to the ocean until the rock and ocean are at the same temperature. Although the ocean warms up ever so slightly, ΔT_{ocean} is so small as to be completely insignificant. For all practical purposes, the ocean is infinite and unchangeable.

An **energy reservoir** is an object or a part of the environment so large that its temperature does not change when heat is transferred between the system and the reservoir. A reservoir at a higher temperature than the system is called a *hot reservoir*.

FIGURE 19.1 Forces $\vec{F}_{gas}$ and $\vec{F}_{ext}$ both do work as the piston moves.

(a)

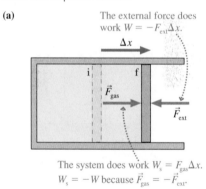

The external force does work $W = -F_{ext}\Delta x$.

The system does work $W_s = F_{gas}\Delta x$. $W_s = -W$ because $\vec{F}_{gas} = -\vec{F}_{ext}$.

(b)

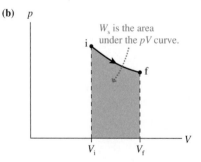

W_s is the area under the pV curve.

FIGURE 19.2 Energy-transfer diagrams.

(a)

Copper bar

Heat is transferred from hot to cold.

Q_H Q_C

Fire

Ice

Hot reservoir at T_H

Cold reservoir at T_C

(b)   Heat energy is transferred from a hot reservoir to a cold reservoir. Energy conservation requires $Q_C = Q_H$.

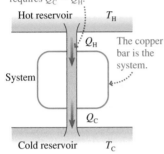

Hot reservoir T_H

Q_H The copper bar is the system.

System

Q_C

Cold reservoir T_C

(c)   The second law forbids a process in which heat is spontaneously transferred from a colder object to a hotter object.

Hot reservoir T_H

Q_H

System

Q_C

Cold reservoir T_C

A vigorously burning flame is a hot reservoir for small objects placed in the flame. A reservoir at a lower temperature than the system is called a *cold reservoir*. The ocean is a cold reservoir for the hot rock. We will use T_H and T_C to designate the temperatures of the hot and cold reservoirs.

Hot and cold reservoirs are idealizations, in the same category as frictionless surfaces and massless strings. No real object can maintain a perfectly constant temperature as heat is transferred in or out. Even so, an object can be modeled as a reservoir if it is much larger than the system that thermally interacts with it.

Heat energy is transferred between a system and a reservoir if they have different temperatures. We will define

$$Q_H = \text{amount of heat transferred to or from a hot reservoir}$$

$$Q_C = \text{amount of heat transferred to or from a cold reservoir}$$

By definition, Q_H and Q_C are *positive* quantities. The direction of heat transfer, which determines the sign of Q in the first law, will always be clear as we deal with thermodynamic devices.

FIGURE 19.2a shows a heavy copper bar between a hot reservoir (at temperature T_H) and a cold reservoir (at temperature T_C). Heat Q_H is transferred from the hot reservoir into the copper and heat Q_C is transferred from the copper to the cold reservoir. FIGURE 19.2b is an **energy-transfer diagram** for this process. The hot reservoir is always drawn at the top, the cold reservoir at the bottom, and the system—the copper bar in this case—between them. Figure 19.2b shows heat Q_H being transferred into the system and Q_C being transferred out.

The first law of thermodynamics $Q = W_s + \Delta E_{th}$ refers to the *system*. Q is the *net* heat to the system, which, in this case, is $Q = Q_H - Q_C$. The copper bar does no work, so $W_s = 0$. The bar warms up when first placed between the two reservoirs, but it soon comes to a steady state where its temperature no longer changes. Then $\Delta E_{th} = 0$. Thus the first law tells us that $Q = Q_H - Q_C = 0$, from which we conclude that $Q_C = Q_H$.

In other words, all of the heat transferred into the hot end of the rod is subsequently transferred out of the cold end. This isn't surprising. After all, we know that heat is transferred spontaneously from a hotter object to a colder object. Even so, there has to be some *means* by which the heat energy gets from the hotter object to the colder. The copper bar provides a route for the energy transfer, and $Q_C = Q_H$ is the statement that energy is conserved as it moves through the bar.

Contrast Figure 19.2b with FIGURE 19.2c. Figure 19.2c shows a system in which heat is being transferred from the cold reservoir to the hot reservoir. The first law of thermodynamics is not violated, because $Q_H = Q_C$, but the second law is. If there were such a system, it would allow the spontaneous (i.e., with no outside input or assistance) transfer of heat from a colder object to a hotter object. The process of Figure 19.2c is forbidden by the second law of thermodynamics.

Work into Heat and Heat into Work

Turning work into heat is easy—just rub two objects together. Work from the friction force increases the objects' thermal energy and their temperature. Heat energy is then transferred from the warmer objects to the cooler environment. FIGURE 19.3 is the energy-transfer diagram for this process. The conversion of work into heat is 100% efficient in that *all* the energy supplied to the system as work is ultimately transferred to the environment as heat. Notice that the objects have returned to their initial state at the end of this process, ready to repeat the process for as long as there's a source of motion.

The reverse—transforming heat into work—isn't so easy. Heat can be transformed into work in a one-time process, such as an isothermal expansion of a gas, but at the end the system is not restored to its initial state. To be practical, **a device that transforms heat into work must return to its initial state at the end of the process and be ready for continued use.** You want your car engine to turn over and over for as long as there's fuel.

FIGURE 19.3 Work can be transformed into heat with 100% efficiency.

Hot reservoir T_H

W System

Q_C

Cold reservoir T_C

Energy-transfer diagrams show work entering or leaving from the side.

Interestingly, no one has ever invented a "perfect engine" that transforms heat into work with 100% efficiency *and returns to its initial state* so that it can continue to do work as long as there is fuel. Of course, that such a device has not been invented is not a proof that it can't be done. We'll provide a proof shortly, but for now we'll make the hypothesis that the process of FIGURE 19.4 is somehow forbidden.

Notice the asymmetry between Figures 19.3 and 19.4. The perfect transformation of work into heat is permitted, but the perfect transformation of heat into work is forbidden. This asymmetry parallels the asymmetry of the two processes in Figure 19.2. In fact, we'll soon see that the "perfect engine" of Figure 19.4 is forbidden for exactly the same reason: the second law of thermodynamics.

FIGURE 19.4 There are no perfect engines that turn heat into work with 100% efficiency.

19.2 Heat Engines and Refrigerators

The steam generator at your local electric power plant works by boiling water to produce high-pressure steam that spins a turbine (which then spins a generator to produce electricity). That is, the steam pressure is doing work. The steam is then condensed to liquid water and pumped back to the boiler to start the process again. There are two crucial ideas here. First, the device works in a cycle, with the water returning to its initial conditions once a cycle. Second, heat is transferred to the water in the boiler, but heat is transferred *out* of the water in the condenser.

Car engines and steam generators are examples of what we call *heat engines*. A **heat engine** is any closed-cycle device that extracts heat Q_H from a hot reservoir, does useful work, and exhausts heat Q_C to a cold reservoir. A **closed-cycle device** is one that periodically *returns to its initial conditions,* repeating the same process over and over. That is, all state variables (pressure, temperature, thermal energy, and so on) return to their initial values once every cycle. Consequently, a heat engine can continue to do useful work for as long as it is attached to the reservoirs.

FIGURE 19.5 is the energy-transfer diagram of a heat engine. Unlike the forbidden "perfect engine" of Figure 19.4, a heat engine is connected to both a hot reservoir *and* a cold reservoir. You can think of a heat engine as "siphoning off" some of the heat that moves from the hot reservoir to the cold reservoir and transforming that heat into work—*some* of the heat, but not all.

Because the temperature and thermal energy of a heat engine return to their initial values at the end of each cycle, there is no *net* change in E_{th}:

$$(\Delta E_{th})_{net} = 0 \quad \text{(any heat engine, over one full cycle)} \quad (19.3)$$

Consequently, the first law of thermodynamics *for a full cycle* of a heat engine is $(\Delta E_{th})_{net} = Q - W_s = 0$.

Let's define W_{out} to be the useful work done *by* the heat engine *per cycle*. The first law applied to a heat engine is

$$W_{out} = Q_{net} = Q_H - Q_C \quad \text{(work per cycle done by a heat engine)} \quad (19.4)$$

This is just energy conservation. The energy-transfer diagram of Figure 19.5 is a pictorial representation of Equation 19.4.

NOTE ▶ Equations 19.3 and 19.4 apply only to a *full cycle* of the heat engine. They are *not* valid for any of the individual processes that make up a cycle. ◀

For practical reasons, we would like an engine to do the maximum amount of work with the minimum amount of fuel. We can measure the performance of a heat engine in terms of its **thermal efficiency** η (lowercase Greek eta), defined as

$$\eta = \frac{W_{out}}{Q_H} = \frac{\text{what you get}}{\text{what you had to pay}} \quad (19.5)$$

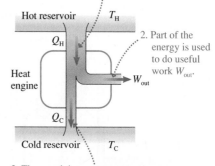

The steam turbine in a modern power plant is an enormous device. Expanding steam does work by spinning the turbine.

FIGURE 19.5 The energy-transfer diagram of a heat engine.

1. Heat energy Q_H is transferred from the hot reservoir (typically burning fuel) to the system.

2. Part of the energy is used to do useful work W_{out}.

3. The remaining energy $Q_C = Q_H - W_{out}$ is exhausted to the cold reservoir (cooling water or the air) as waste heat.

FIGURE 19.6 η is the fraction of heat energy that is transformed into useful work.

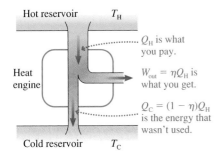

Using Equation 19.4 for W_{out}, we can also write the thermal efficiency as

$$\eta = 1 - \frac{Q_C}{Q_H} \qquad (19.6)$$

FIGURE 19.6 illustrates the idea of thermal efficiency.

A *perfect* heat engine would have $\eta_{perfect} = 1$. That is, it would be 100% efficient at converting heat from the hot reservoir (the burning fuel) into work. You can see from Equation 19.6 that a perfect engine would have no exhaust ($Q_C = 0$) and would not need a cold reservoir. Figure 19.4 has already suggested that there are no perfect heat engines, that an engine with $\eta = 1$ is impossible. A heat engine *must* exhaust **waste heat** to a cold reservoir. It is energy that was extracted from the hot reservoir but *not* transformed to useful work.

Practical heat engines, such as car engines and steam generators, have thermal efficiencies in the range $\eta \approx 0.1 - 0.5$. This is not large. Can a clever designer do better, or is this some kind of physical limitation?

STOP TO THINK 19.1 Rank in order, from largest to smallest, the work W_{out} performed by these four heat engines.

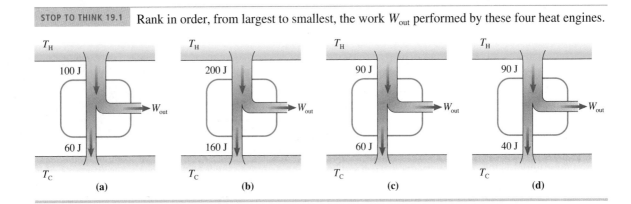

A Heat-Engine Example

To illustrate how these ideas actually work, FIGURE 19.7 shows a simple engine that converts heat into the work of lifting mass M.

FIGURE 19.7 A simple heat engine transforms heat into work.

(a) Heat is transferred into the gas from the burning fuel.

(b) The gas does work by lifting the mass in an isobaric expansion.

(c) The piston is locked and the mass is removed. The heat is turned off.

(d) The gas cools back to room temperature at constant volume. Then the piston is unlocked.

(e) A steadily increasing external force steadily raises the pressure in an isothermal compression until the pressure has been restored to its initial value.

Isobaric heating and expansion   **Constant-volume cooling**   **Isothermal compression**

The net effect of this multistep process is to convert some of the fuel's energy into the useful work of lifting the mass. There has been no net change in the gas, which has returned to its initial pressure, volume, and temperature at the end of step (e). We can start the whole process over again and continue lifting masses (doing work) as long as we have fuel.

FIGURE 19.8 shows the heat-engine process on a pV diagram. It is a *closed cycle* because the gas returns to its initial conditions. No work is done during the isochoric process, and, as you can see from the areas under the curve, the work done *by* the gas to lift the mass is greater than the work the environment must do *on* the gas to recompress it. Thus this heat engine, by burning fuel, does *net* work per cycle:
$W_{net} = W_{lift} - W_{ext} = (W_s)_{1\rightarrow 2} + (W_s)_{3\rightarrow 1}$.

Notice that the cyclical process of Figure 19.8 involves two cooling processes in which heat is transferred *from* the gas to the environment. Heat energy is transferred from hotter objects to colder objects, so the system *must* be connected to a cold reservoir with $T_C < T_{gas}$ during these two processes. A key to understanding heat engines is that they require both a heat source (burning fuel) *and* a heat sink (cooling water, the air, or something at a lower temperature than the system).

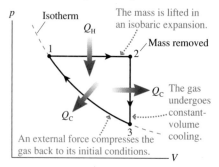

FIGURE 19.8 The closed-cycle pV diagram for the heat engine of Figure 19.7.

EXAMPLE 19.1 Analyzing a heat engine I

Analyze the heat engine of FIGURE 19.9 to determine (a) the net work done per cycle, (b) the engine's thermal efficiency, and (c) the engine's power output if it runs at 600 rpm. Assume the gas is monatomic.

FIGURE 19.9 The heat engine of Example 19.1.

MODEL The gas follows a closed cycle consisting of three distinct processes, each of which was studied in Chapters 16 and 17. For each of the three we need to determine the work done and the heat transferred.

SOLVE To begin, we can use the initial conditions at state 1 and the ideal-gas law to determine the number of moles of gas:

$$n = \frac{p_1 V_1}{RT_1} = \frac{(200 \times 10^3 \text{ Pa})(2.0 \times 10^{-4} \text{ m}^3)}{(8.31 \text{ J/mol K})(300 \text{ K})} = 0.0160 \text{ mol}$$

Process 1 → 2: The work done *by* the gas in the isobaric expansion is

$$(W_s)_{12} = p\Delta V = (200 \times 10^3 \text{ Pa})\big((6.0 - 2.0) \times 10^{-4} \text{ m}^3\big) = 80 \text{ J}$$

We can use the ideal-gas law at constant pressure to find $T_2 = (V_2/V_1)T_1 = 3T_1 = 900$ K. The heat transfer during a constant-pressure process is

$$Q_{12} = nC_P\Delta T$$
$$= (0.0160 \text{ mol})(20.8 \text{ J/mol K})(900 \text{ K} - 300 \text{ K}) = 200 \text{ J}$$

where we used $C_P = \frac{5}{2}R$ for a monatomic ideal gas.

Process 2 → 3: No work is done in an isochoric process, so $(W_s)_{23} = 0$. The temperature drops back to 300 K, so the heat transfer, with $C_V = \frac{3}{2}R$, is

$$Q_{23} = nC_V\Delta T$$
$$= (0.0160 \text{ mol})(12.5 \text{ J/mol K})(300 \text{ K} - 900 \text{ K}) = -120 \text{ J}$$

Process 3 → 1: The gas returns to its initial state with volume V_1. The work done *by* the gas during an isothermal process is

$$(W_s)_{31} = nRT \ln\left(\frac{V_1}{V_3}\right)$$
$$= (0.0160 \text{ mol})(8.31 \text{ J/mol K})(300 \text{ K}) \ln\left(\frac{1}{3}\right) = -44 \text{ J}$$

W_s is negative because the environment does work on the gas to compress it. An isothermal process has $\Delta E_{th} = 0$ and hence, from the first law,

$$Q_{31} = (W_s)_{31} = -44 \text{ J}$$

Q is negative because the gas must be cooled as it is compressed to keep the temperature constant.

a. The *net* work done by the engine during one cycle is

$$W_{out} = (W_s)_{12} + (W_s)_{23} + (W_s)_{31} = 36 \text{ J}$$

As a consistency check, notice that the net heat transfer is

$$Q_{net} = Q_{12} + Q_{23} + Q_{31} = 36 \text{ J}$$

Equation 19.4 told us that a heat engine *must* have $W_{out} = Q_{net}$, and we see that it does.

b. The efficiency depends not on the net heat transfer but on the heat Q_H transferred into the engine from the flame. Heat enters during process 1 → 2, where Q is positive, and exits during processes 2 → 3 and 3 → 1, where Q is negative. Thus

$$Q_H = Q_{12} = 200 \text{ J}$$
$$Q_C = |Q_{23}| + |Q_{31}| = 164 \text{ J}$$

Notice that $Q_H - Q_C = 36$ J $= W_{out}$. In this heat engine, 200 J of heat from the hot reservoir does 36 J of useful work. Thus the thermal efficiency is

$$\eta = \frac{W_{out}}{Q_H} = \frac{36 \text{ J}}{200 \text{ J}} = 0.18 \text{ or } 18\%$$

This heat engine is far from being a perfect engine!

Continued

c. An engine running at 600 rpm goes through 10 cycles per second. The power output is the work done *per second:*

$$P_{out} = (\text{work per cycle}) \times (\text{cycles per second})$$
$$= 360 \text{ J/s} = 360 \text{ W}$$

ASSESS Although we didn't need Q_{net}, verifying that $Q_{net} = W_{out}$ was a check of self-consistency. Heat-engine analysis requires many calculations and offers many opportunities to get signs wrong. However, there are a sufficient number of self-consistency checks so that you can almost always spot calculational errors *if you check for them.*

Let's think about this example a bit more before going on. We've said that a heat engine operates between a hot reservoir and a cold reservoir. Figure 19.9 doesn't explicitly show the reservoirs. Nonetheless, we know that heat is transferred from a hotter object to a colder object. Heat Q_H is transferred into the system during process $1 \rightarrow 2$ as the gas warms from 300 K to 900 K. For this to be true, the hot-reservoir temperature T_H must be $\geq$900 K. Likewise, heat Q_C is transferred from the system to the cold reservoir as the temperature drops from 900 K to 300 K in process $2 \rightarrow 3$. For this to be true, the cold-reservoir temperature T_C must be $\leq$300 K.

So, while we really don't know what the reservoirs are or their exact temperatures, we can say with certainty that the hot-reservoir temperature T_H must exceed the highest temperature reached by the system and the cold-reservoir temperature T_C must be less than the coldest system temperature.

Refrigerators

Your house or apartment has a refrigerator. Very likely it has an air conditioner. The purpose of these devices is to make air that is cooler than its environment even colder. The first does so by blowing hot air out into a warm room, the second by blowing it out to the hot outdoors. You've probably felt the hot air exhausted by an air conditioner compressor or coming out from beneath the refrigerator.

At first glance, a refrigerator or air conditioner may seem to violate the second law of thermodynamics. After all, doesn't the second law forbid heat from being transferred from a colder object to a hotter object? Not quite: The second law says that heat is not *spontaneously* transferred from a colder to a hotter object. A refrigerator or air conditioner requires electric power to operate. They do cause heat to be transferred from cold to hot, but the transfer is "assisted" rather than spontaneous.

A **refrigerator** is any closed-cycle device that uses external work W_{in} to remove heat Q_C from a cold reservoir and exhaust heat Q_H to a hot reservoir. **FIGURE 19.10** is the energy-transfer diagram of a refrigerator. The cold reservoir is the air inside the refrigerator or the air inside your house on a summer day. To keep the air cold, in the face of inevitable "heat leaks," the refrigerator or air conditioner compressor continuously removes heat from the cold reservoir and exhausts heat into the room or outdoors. You can think of a refrigerator as "pumping heat uphill," much as a water pump lifts water uphill.

Because a refrigerator, like a heat engine, is a cyclical device, $\Delta E_{th} = 0$. Conservation of energy requires

$$Q_H = Q_C + W_{in} \tag{19.7}$$

To move energy from a colder to a hotter reservoir, a refrigerator must exhaust *more* heat to the outside than it removes from the inside. This has significant implications for whether or not you can cool a room by leaving the refrigerator door open.

The thermal efficiency of a heat engine was defined as "what you get (useful work W_{out})" versus "what you had to pay (fuel to supply Q_H)." By analogy, we define the **coefficient of performance** K of a refrigerator to be

$$K = \frac{Q_C}{W_{in}} = \frac{\text{what you get}}{\text{what you had to pay}} \tag{19.8}$$

What you get, in this case, is the removal of heat from the cold reservoir. But you have to pay the electric company for the work needed to run the refrigerator. A better

This air conditioner transfers heat energy *from* the cool indoors *to* the hot exterior.

FIGURE 19.10 The energy-transfer diagram of a refrigerator.

The amount of heat exhausted to the hot reservoir is larger than the amount of heat extracted from the cold reservoir.

External work is used to remove heat from a cold reservoir and exhaust heat to a hot reservoir.

refrigerator will require less work to remove a given amount of heat, thus having a larger coefficient of performance.

A perfect refrigerator would require no work ($W_{in} = 0$) and would have $K_{perfect} = \infty$. But if Figure 19.10 had no work input, it would look like Figure 19.2c. That device was forbidden by the second law of thermodynamics because, with no work input, heat would move *spontaneously* from cold to hot.

We noted in Chapter 18 that the second law of thermodynamics can be stated several different but equivalent ways. We can now give a third statement:

> **Second law, informal statement #3** There are no perfect refrigerators with coefficient of performance $K = \infty$.

Any real refrigerator or air conditioner *must* use work to move energy from the cold reservoir to the hot reservoir, hence $K < \infty$.

No Perfect Heat Engines

We hypothesized above that there are no perfect heat engines—that is, no heat engines like the one shown in Figure 19.4 with $Q_C = 0$ and $\eta = 1$. Now we're ready to prove this hypothesis. **FIGURE 19.11** shows a hot reservoir at temperature T_H and a cold reservoir at temperature T_C. An ordinary refrigerator, one that obeys all the laws of physics, is operating between these two reservoirs.

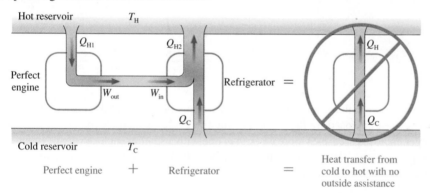

FIGURE 19.11 A perfect engine driving an ordinary refrigerator would be able to violate the second law of thermodynamics.

Suppose we had a perfect heat engine, one that takes in heat Q_H from the high-temperature reservoir and transforms that energy entirely into work W_{out}. If we had such a heat engine, we could use its output to provide the work input to the refrigerator. The two devices combined have no connection to the external world. That is, there's no net input or net output of work.

If we built a box around the heat engine and refrigerator, so that you couldn't see what was inside, the only thing you would observe is heat being transferred *with no outside assistance* from the cold reservoir to the hot reservoir. But a spontaneous or unassisted transfer of heat from a colder to a hotter object is exactly what the second law of thermodynamics forbids. Consequently, our assumption of a perfect heat engine must be wrong. Hence another statement of the second law of thermodynamics is:

> **Second law, informal statement #4** There are no perfect heat engines with efficiency $\eta = 1$.

Any real heat engine *must* exhaust waste heat Q_C to a cold reservoir.

STOP TO THINK 19.2 It's a hot day and your air conditioner is broken. Your roommate says, "Let's open the refrigerator door and cool this place off." Will this work?

a. Yes.　　b. No.　　c. It might, but it will depend on how hot the room is.

19.3 Ideal-Gas Heat Engines

We will focus on heat engines that use a gas as the *working substance*. The gasoline or diesel engine in your car is an engine that alternately compresses and expands a gaseous fuel-air mixture. A discussion of engines such as steam generators that rely on phase changes will be deferred to more advanced courses.

A gas heat engine can be represented by a closed-cycle trajectory in the *pV* diagram, such as the one shown in FIGURE 19.12a. This observation leads to an important geometric interpretation of the work done by the system during one full cycle. You learned in Section 19.1 that the work done *by* the system is the area under the curve of a *pV* trajectory. As FIGURE 19.12b shows, the net work done during a full cycle is

$$W_{\text{out}} = W_{\text{expand}} - |W_{\text{compress}}| = \text{area } \textit{inside} \text{ the closed curve} \qquad (19.9)$$

FIGURE 19.12 The work W_{out} done by the system during one full cycle is the area enclosed within the curve.

(a)

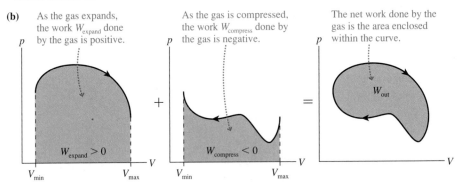

(b)

You can see that **the net work done by a gas heat engine during one full cycle is the area enclosed by the *pV* curve for the cycle.** A thermodynamic cycle with a larger enclosed area does more work than one with a smaller enclosed area. Notice that the gas must go around the *pV* trajectory in a *clockwise* direction for W_{out} to be positive. We'll see later that a refrigerator uses a counterclockwise (ccw) cycle.

Ideal-Gas Summary

We've learned a lot about ideal gases in the last three chapters. All gas processes obey the ideal-gas law $pV = nRT$ and the first law of thermodynamics $\Delta E_{\text{th}} = Q - W_{\text{s}}$. Table 19.1 summarizes the results for specific gas processes. This table shows W_{s}, the work done *by* the system, so the signs are opposite those in Chapter 17.

TABLE 19.1 Summary of ideal-gas processes

Process	Gas law	Work W_{s}	Heat Q	Thermal energy
Isochoric	$p_i/T_i = p_f/T_f$	0	$nC_V \Delta T$	$\Delta E_{\text{th}} = Q$
Isobaric	$V_i/T_i = V_f/T_f$	$p\Delta V$	$nC_P \Delta T$	$\Delta E_{\text{th}} = Q - W_{\text{s}}$
Isothermal	$p_iV_i = p_fV_f$	$nRT \ln(V_f/V_i)$ $pV \ln(V_f/V_i)$	$Q = W_{\text{s}}$	$\Delta E_{\text{th}} = 0$
Adiabatic	$p_iV_i^\gamma = p_fV_f^\gamma$ $T_iV_i^{\gamma-1} = T_fV_f^{\gamma-1}$	$(p_fV_f - p_iV_i)/(1-\gamma)$ $-nC_V \Delta T$	0	$\Delta E_{\text{th}} = -W_{\text{s}}$
Any	$p_iV_i/T_i = p_fV_f/T_f$	area under curve		$\Delta E_{\text{th}} = nC_V \Delta T$

There is one entry in this table that you haven't seen before. The expression

$$W_{\text{s}} = \frac{p_fV_f - p_iV_i}{1 - \gamma} \qquad \text{(work in an adiabatic process)} \qquad (19.10)$$

for the work done in an adiabatic process follows from writing $W_s = -\Delta E_{th} = -nC_V \Delta T$, which you learned in Chapter 17, then using $\Delta T = \Delta(pV)/nR$ and the definition of γ. The proof will be left for a homework problem.

You learned in Chapter 18 that the thermal energy of an ideal gas depends only on its temperature. Table 19.2 lists the thermal energy, molar specific heats, and specific heat ratio $\gamma = C_P/C_V$ for monatomic and diatomic gases.

A Strategy for Heat-Engine Problems

The engine of Example 19.1 was not a realistic heat engine, but it did illustrate the kinds of reasoning and computations involved in the analysis of a heat engine.

TABLE 19.2 Properties of monatomic and diatomic gases

	Monatomic	Diatomic
E_{th}	$\frac{3}{2}nRT$	$\frac{5}{2}nRT$
C_V	$\frac{3}{2}R$	$\frac{5}{2}R$
C_P	$\frac{5}{2}R$	$\frac{7}{2}R$
γ	$\frac{5}{3} = 1.67$	$\frac{7}{5} = 1.40$

PROBLEM-SOLVING
STRATEGY 19.1   **Heat-engine problems**   (MP)

MODEL Identify each process in the cycle.

VISUALIZE Draw the pV diagram of the cycle.

SOLVE There are several steps in the mathematical analysis.

- Use the ideal-gas law to complete your knowledge of n, p, V, and T at one point in the cycle.
- Use the ideal-gas law and equations for specific gas processes to determine p, V, and T at the beginning and end of each process.
- Calculate Q, W_s, and ΔE_{th} for each process.
- Find W_{out} by adding W_s for each process in the cycle. If the geometry is simple, you can confirm this value by finding the area enclosed within the pV curve.
- Add just the *positive* values of Q to find Q_H.
- Verify that $(\Delta E_{th})_{net} = 0$. This is a self-consistency check to verify that you haven't made any mistakes.
- Calculate the thermal efficiency η and any other quantities you need to complete the solution.

ASSESS Is $(\Delta E_{th})_{net} = 0$? Do all the signs of W_s and Q make sense? Does η have a reasonable value? Have you answered the question?

EXAMPLE 19.2   **Analyzing a heat engine II**

A heat engine with a diatomic gas as the working substance uses the closed cycle shown in FIGURE 19.13. How much work does this engine do per cycle, and what is its thermal efficiency?

FIGURE 19.13 The pV diagram for the heat engine of Example 19.2.

MODEL Processes $1 \rightarrow 2$ and $3 \rightarrow 4$ are isobaric. Processes $2 \rightarrow 3$ and $4 \rightarrow 1$ are isochoric.

VISUALIZE The pV diagram has already been drawn.

SOLVE We know the pressure, volume, and temperature at state 4. The number of moles of gas in the heat engine is

$$n = \frac{p_4 V_4}{RT_4} = \frac{(101,300 \text{ Pa})(1.0 \text{ m}^3)}{(8.31 \text{ J/mol K})(300 \text{ K})} = 40.6 \text{ mol}$$

p/T = constant during an isochoric process and V/T = constant during an isobaric process. These allow us to find that $T_1 = T_3 = 900$ K and $T_2 = 2700$ K. This completes our knowledge of the state variables at all four corners of the diagram.

Process $1 \rightarrow 2$ is an isobaric expansion, so

$$(W_s)_{12} = p\Delta V = (3.0 \times 101,300 \text{ Pa})(2.0 \text{ m}^3) = 6.08 \times 10^5 \text{ J}$$

Continued

where we converted the pressure to pascals. The heat transfer during an isobaric expansion is

$$Q_{12} = nC_P \Delta T = (40.6 \text{ mol})(29.1 \text{ J/mol K})(1800 \text{ K})$$

$$= 21.27 \times 10^5 \text{ J}$$

where $C_P = \frac{7}{2}R$ for a diatomic gas. Then, using the first law,

$$\Delta E_{12} = Q_{12} - (W_s)_{12} = 15.19 \times 10^5 \text{ J}$$

Process $2 \rightarrow 3$ is an isochoric process, so $(W_s)_{23} = 0$ and

$$\Delta E_{23} = Q_{23} = nC_V \Delta T = -15.19 \times 10^5 \text{ J}$$

Notice that ΔT is *negative*.

Process $3 \rightarrow 4$ is an isobaric compression. Now ΔV is negative, so

$$(W_s)_{34} = p\Delta V = -2.03 \times 10^5 \text{ J}$$

and

$$Q_{34} = nC_P \Delta T = -7.09 \times 10^5 \text{ J}$$

Then $\Delta E_{th} = Q_{34} - (W_s)_{34} = -5.06 \times 10^5 \text{ J}$.

Process $4 \rightarrow 1$ is another constant-volume process, so again $(W_s)_{41} = 0$ and

$$\Delta E_{41} = Q_{41} = nC_V \Delta T = 5.06 \times 10^5 \text{ J}$$

The results of all four processes are shown in Table 19.3. The net results for W_{out}, Q_{net}, and $(\Delta E_{th})_{net}$ are found by summing the columns. As expected, $W_{out} = Q_{net}$ and $(\Delta E_{th})_{net} = 0$.

TABLE 19.3 Energy transfers in Example 19.2. All energies $\times 10^5$ J

Process	W_s	Q	ΔE_{th}
$1 \rightarrow 2$	6.08	21.27	15.19
$2 \rightarrow 3$	0	−15.19	−15.19
$3 \rightarrow 4$	−2.03	−7.09	−5.06
$4 \rightarrow 1$	0	5.06	5.06
Net	4.05	4.05	0

The work done during one cycle is $W_{out} = 4.05 \times 10^5$ J. Heat enters the system from the hot reservoir during processes $1 \rightarrow 2$ and $4 \rightarrow 1$, where Q is positive. Summing these gives $Q_H = 26.33 \times 10^5$ J. Thus the thermal efficiency of this engine is

$$\eta = \frac{W_{out}}{Q_H} = \frac{4.05 \times 10^5 \text{ J}}{26.33 \times 10^5 \text{ J}} = 0.15 = 15\%$$

ASSESS The verification that $W_{out} = Q_{net}$ and $(\Delta E_{th})_{net} = 0$ gives us great confidence that we didn't make any calculational errors. This engine may not seem very efficient, but η is quite typical of many real engines.

We noted in Example 19.1 that a heat engine's hot-reservoir temperature T_H must exceed the highest temperature reached by the system and the cold-reservoir temperature T_C must be less than the coldest system temperature. Although we don't know what the reservoirs are in Example 19.2, we can be sure that $T_H > 2700$ K and $T_C < 300$ K.

STOP TO THINK 19.3 What is the thermal efficiency of this heat engine?

a. 0.10
b. 0.50
c. 0.25
d. 4
e. Can't tell without knowing Q_C

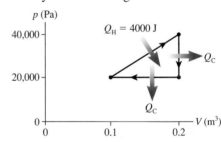

The Brayton Cycle

The heat engines of Examples 19.1 and 19.2 have been educational but not realistic. As an example of a more realistic heat engine we'll look at the thermodynamic cycle known as the *Brayton cycle*. It is a reasonable model of a *gas turbine engine*. Gas turbines are used for electric power generation and as the basis for jet engines in aircraft and rockets. The *Otto cycle,* which describes the gasoline internal combustion engine, and the *Diesel cycle,* which, not surprisingly, describes the diesel engine, will be the subject of homework problems.

FIGURE 19.14a is a schematic look at a gas turbine engine, and FIGURE 19.14b is the corresponding pV diagram. To begin the Brayton cycle, air at an initial pressure p_1 is rapidly compressed in a *compressor*. This is an *adiabatic process*, with $Q = 0$,

A jet engine uses a modified Brayton cycle.

because there is no time for heat to be exchanged with the surroundings. Recall that an adiabatic compression raises the temperature of a gas by doing work on it, not by heating it, so the air leaving the compressor is very hot.

The hot gas flows into a combustion chamber. Fuel is continuously admitted to the combustion chamber where it mixes with the hot gas and is ignited, transferring heat to the gas at constant pressure and raising the gas temperature yet further. The high-pressure gas then expands, spinning a turbine that does some form of useful work. This adiabatic expansion, with $Q = 0$, drops the temperature and pressure of the gas. The pressure at the end of the expansion through the turbine is back to p_1, but the gas is still quite hot. The gas completes the cycle by flowing through a device called a **heat exchanger** that transfers heat energy to a cooling fluid. Large power plants are often sited on rivers or oceans in order to use the water for the cooling fluid in the heat exchanger.

This thermodynamic cycle, called a Brayton cycle, has two adiabatic processes—the compression and the expansion through the turbine—plus a constant-pressure heating and a constant-pressure cooling. There's no heat transfer during the adiabatic processes. The hot-reservoir temperature must be $T_H \geq T_3$ for heat to be transferred into the gas during process $2 \to 3$. Similarly, the heat exchanger will remove heat from the gas only if $T_C \leq T_1$.

The thermal efficiency of any heat engine is

$$\eta = \frac{W_{out}}{Q_H} = 1 - \frac{Q_C}{Q_H}$$

Heat is transferred into the gas only during process $2 \to 3$. This is an isobaric process, so $Q_H = nC_P\Delta T = nC_P(T_3 - T_2)$. Similarly, heat is transferred out only during the isobaric process $4 \to 1$.

We have to be careful with signs. Q_{41} is negative because the temperature decreases, but Q_C was defined as the *amount* of heat exchanged with the cold reservoir, a positive quantity. Thus

$$Q_C = |Q_{41}| = |nC_P(T_1 - T_4)| = nC_P(T_4 - T_1) \qquad (19.11)$$

With these expressions for Q_H and Q_C, the thermal efficiency is

$$\eta_{Brayton} = 1 - \frac{T_4 - T_1}{T_3 - T_2} \qquad (19.12)$$

This expression isn't useful unless we compute all four temperatures. Fortunately, we can cast Equation 19.12 into a more useful form.

You learned in Chapter 17 that $pV^\gamma = $ constant during an adiabatic process, where $\gamma = C_P/C_V$ is the specific heat ratio. If we use $V = nRT/p$ from the ideal-gas law, $V^\gamma = (nR)^\gamma T^\gamma p^{-\gamma}$. $(nR)^\gamma$ is a constant, so we can write $pV^\gamma = $ constant as

$$p^{(1-\gamma)}T^\gamma = \text{constant} \qquad (19.13)$$

Equation 19.13 is a pressure-temperature relationship for an adiabatic process. Because $(T^\gamma)^{1/\gamma} = T$, we can simplify Equation 19.13 by raising both sides to the power $1/\gamma$. Doing so gives

$$p^{(1-\gamma)/\gamma}T = \text{constant} \qquad (19.14)$$

during an adiabatic process.

Process $1 \to 2$ is an adiabatic process; hence

$$p_1^{(1-\gamma)/\gamma}T_1 = p_2^{(1-\gamma)/\gamma}T_2 \qquad (19.15)$$

Isolating T_1 gives

$$T_1 = \frac{p_2^{(1-\gamma)/\gamma}}{p_1^{(1-\gamma)/\gamma}}T_2 = \left(\frac{p_2}{p_1}\right)^{(1-\gamma)/\gamma} T_2 = \left(\frac{p_{max}}{p_{min}}\right)^{(1-\gamma)/\gamma} T_2 \qquad (19.16)$$

If we define the **pressure ratio** r_p as $r_p = p_{max}/p_{min}$, then T_1 and T_2 are related by

$$T_1 = r_p^{(1-\gamma)/\gamma}T_2 \qquad (19.17)$$

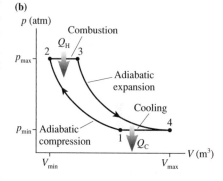

FIGURE 19.14 A gas turbine engine follows a Brayton cycle.

FIGURE 19.15 The efficiency of a Brayton cycle as a function of the pressure ratio r_p.

The efficiency first grows quickly as the pressure ratio is increased, reaching ≈ 50% at $r_p = 10$, then levels off.

Any increase in efficiency beyond ≈ 50% has to be weighed against the higher costs of a better compressor that can achieve a much higher pressure ratio.

The algebra of getting to Equation 19.17 was a bit tricky, but the final result is fairly simple.

Process $3 \rightarrow 4$ is also an adiabatic process. The same reasoning leads to

$$T_4 = r_p^{(1-\gamma)/\gamma} T_3 \qquad (19.18)$$

If we substitute these expressions for T_1 and T_4 into Equation 19.12, the efficiency is

$$\eta_B = 1 - \frac{T_4 - T_1}{T_3 - T_2} = 1 - \frac{r_p^{(1-\gamma)/\gamma} T_3 - r_p^{(1-\gamma)/\gamma} T_2}{T_3 - T_2} = 1 - \frac{r_p^{(1-\gamma)/\gamma}(T_3 - T_2)}{T_3 - T_2}$$

$$= 1 - r_p^{(1-\gamma)/\gamma}$$

Remarkably, all the temperatures cancel and we're left with an expression that depends only on the pressure ratio. Noting that $(1 - \gamma)$ is negative, we can make one final change and write

$$\eta_B = 1 - \frac{1}{r_p^{(\gamma-1)/\gamma}} \qquad (19.19)$$

FIGURE 19.15 is a graph of the efficiency of the Brayton cycle as a function of the pressure ratio, assuming $\gamma = 1.40$ for a diatomic gas such as air.

19.4 Ideal-Gas Refrigerators

Suppose we were to operate a Brayton heat engine backward, going ccw rather than cw in the pV diagram. FIGURE 19.16a, (which you should compare to Figure 19.14a) shows a device for doing this. FIGURE 19.16b is its pV diagram, and FIGURE 19.16c is the energy-transfer diagram. Starting from point 4, the gas is adiabatically compressed to increase its temperature and pressure. It then flows through a high-temperature heat exchanger where the gas *cools* at constant pressure from temperature T_3 to T_2. The gas then expands adiabatically, leaving it significantly colder at T_1 than it started at T_4. It completes the cycle by flowing through a low-temperature heat exchanger, where it *warms* back to its starting temperature.

FIGURE 19.16 A refrigerator that extracts heat from the cold reservoir and exhausts heat to the hot reservoir.

(a)

(b)

(c)

Suppose that the low-temperature heat exchanger is a closed container of air surrounding a pipe through which the engine's cold gas is flowing. The heat-exchange process $1 \rightarrow 4$ *cools* the air in the container as it warms the gas flowing through the pipe. If you were to place eggs and milk inside this closed container, you would call it a refrigerator!

Going around a closed pV cycle in a ccw direction reverses the sign of W for each process in the cycle. Consequently, the area inside the curve of Figure 19.16b is W_{in}, the work done *on* the system. Here work is used to extract heat Q_C from the cold reservoir and exhaust a larger amount of heat $Q_H = Q_C + W_{in}$ to the hot reservoir. But where, in this situation, are the energy reservoirs?

Understanding a refrigerator is a little harder than understanding a heat engine. The key is to remember that **heat is always transferred from a hotter object to a colder object**. In particular,

- The gas in a refrigerator can extract heat Q_C *from* the cold reservoir only if the gas temperature is *lower* than the cold-reservoir temperature T_C. Heat energy is then transferred *from* the cold reservoir *into* the colder gas.
- The gas in a refrigerator can exhaust heat Q_H *to* the hot reservoir only if the gas temperature is *higher* than the hot-reservoir temperature T_H. Heat energy is then transferred *from* the warmer gas *into* the hot reservoir.

These two requirements place severe constraints on the thermodynamics of a refrigerator. Because there is no reservoir colder than T_C, the gas cannot reach a temperature lower than T_C by heat exchange. The gas in a refrigerator *must* use an adiabatic expansion ($Q = 0$) to lower the temperature below T_C. Likewise, a gas refrigerator requires an adiabatic compression to raise the gas temperature above T_H.

EXAMPLE 19.3 **Analyzing a refrigerator**

A refrigerator using helium gas operates on a reversed Brayton cycle with a pressure ratio of 5.0. Prior to compression, the gas occupies 100 cm^3 at a pressure of 150 kPa and a temperature of $-23°C$. Its volume at the end of the expansion is 80 cm^3. What are the refrigerator's coefficient of performance and its power input if it operates at 60 cycles per second?

MODEL The Brayton cycle has two adiabatic processes and two isobaric processes. The work per cycle needed to run the refrigerator is $W_{in} = Q_H - Q_C$; hence we can determine both the coefficient of performance and the power requirements from Q_H and Q_C. Heat energy is transferred only during the two isobaric processes.

VISUALIZE FIGURE 19.17 shows the pV cycle. We know from the pressure ratio of 5.0 that the maximum pressure is 750 kPa. Neither V_2 nor V_3 is known.

FIGURE 19.17 A Brayton-cycle refrigerator.

SOLVE To calculate heat we're going to need the temperatures at the four corners of the cycle. First, we can use the conditions of state 4 to find the number of moles of helium:

$$n = \frac{p_4 V_4}{R T_4} = 0.00722 \text{ mol}$$

Process $1 \rightarrow 4$ is isobaric; hence temperature T_1 is

$$T_1 = \frac{V_1}{V_4} T_4 = (0.80)(250 \text{ K}) = 200 \text{ K} = -73°C$$

With Equation 19.14 we found that the quantity $p^{(1-\gamma)/\gamma} T$ remains constant during an adiabatic process. Helium is a monatomic gas with $\gamma = \frac{5}{3}$, so $(1 - \gamma)/\gamma = -\frac{2}{5} = -0.40$. For the adiabatic compression $4 \rightarrow 3$,

$$p_3^{-0.40} T_3 = p_4^{-0.40} T_4$$

Solving for T_3 gives

$$T_3 = \left(\frac{p_4}{p_3}\right)^{-0.40} T_4 = \left(\frac{1}{5}\right)^{-0.40} (250 \text{ K}) = 476 \text{ K} = 203°C$$

The same analysis applied to the $2 \rightarrow 1$ adiabatic expansion gives

$$T_2 = \left(\frac{p_1}{p_2}\right)^{-0.40} T_1 = \left(\frac{1}{5}\right)^{-0.40} (200 \text{ K}) = 381 \text{ K} = 108°C$$

Now we can use $C_P = \frac{5}{2}R = 20.8 \text{ J/mol K}$ for a monatomic gas to compute the heat transfers:

$$Q_H = |Q_{32}| = nC_P(T_3 - T_2)$$
$$= (0.00722 \text{ mol})(20.8 \text{ J/mol K})(95 \text{ K}) = 14.3 \text{ J}$$
$$Q_C = |Q_{14}| = nC_P(T_4 - T_1)$$
$$= (0.00722 \text{ mol})(20.8 \text{ J/mol K})(50 \text{ K}) = 7.5 \text{ J}$$

Thus the work *input* to the refrigerator is $W_{in} = Q_H - Q_C = 6.8 \text{ J}$. During each cycle, 6.8 J of work are done *on* the gas to extract 7.5 J of heat from the cold reservoir. Then 14.3 J of heat are exhausted into the hot reservoir.

The refrigerator's coefficient of performance is

$$K = \frac{Q_C}{W_{in}} = \frac{7.5 \text{ J}}{6.8 \text{ J}} = 1.1$$

The power input needed to run the refrigerator is

$$P_{in} = 6.8 \frac{\text{J}}{\text{cycle}} \times 60 \frac{\text{cycles}}{\text{s}} = 410 \frac{\text{J}}{\text{s}} = 410 \text{ W}$$

ASSESS These are fairly realistic values for a kitchen refrigerator. You pay your electric company for providing the work W_{in} that operates the refrigerator. The cold reservoir is the freezer compartment. The cold temperature T_C must be higher than T_4 ($T_C > -23°C$) in order for heat to be transferred *from* the cold reservoir *to* the gas. A typical freezer temperature is $-15°C$, so this condition is satisfied. The hot reservoir is the air in the room. The back and underside of a refrigerator have heat-exchanger coils where the hot gas, after compression, transfers heat to the air. The hot temperature T_H must be less than T_2 ($T_H < 108°C$) in order for heat to be transferred *from* the gas *to* the air. An air temperature $\approx 25°C$ under a refrigerator satisfies this condition.

STOP TO THINK 19.4 What, if anything, is wrong with this refrigerator?

19.5 The Limits of Efficiency

Thermodynamics has its historical roots in the development of the steam engine and other machines of the early industrial revolution. Early steam engines, built on the basis of experience rather than scientific understanding, were not very efficient at converting fuel energy into work. The first major theoretical analysis of heat engines was published by the French engineer Sadi Carnot in 1824. The question that Carnot raised was: Can we make a heat engine whose thermal efficiency η approaches 1, or is there an upper limit η_{max} that cannot be exceeded? To frame the question more clearly, imagine we have a hot reservoir at temperature T_H and a cold reservoir at T_C. What is the most efficient heat engine (maximum η) that can operate between these two energy reservoirs? Similarly, what is the most efficient refrigerator (maximum K) that can operate between the two reservoirs?

We just saw that a refrigerator is, in some sense, a heat engine running backward. We might thus suspect that the most efficient heat engine is related to the most efficient refrigerator. Suppose we have a heat engine that we can turn into a refrigerator by reversing the direction of operation, thus changing the direction of the energy transfers, and with *no other changes*. In particular, the heat engine and the refrigerator operate between the same two energy reservoirs at temperatures T_H and T_C.

FIGURE 19.18a shows such a heat engine and its corresponding refrigerator. Notice that the refrigerator has *exactly the same* work and heat transfer as the heat engine, only in the opposite directions. A device that can be operated as either a heat engine or a refrigerator between the same two energy reservoirs and with the same energy transfers, with only their direction changed, is called a **perfectly reversible engine.** A perfectly reversible engine is an idealization, as was the concept of a perfectly elastic collision. Nonetheless, it will allow us to establish limits that no real engine can exceed.

FIGURE 19.18 If a perfectly reversible heat engine is used to operate a perfectly reversible refrigerator, the two devices exactly cancel each other.

Suppose we have a perfectly reversible heat engine and a perfectly reversible refrigerator (the same device running backward) operating between a hot reservoir at temperature T_H and a cold reservoir at temperature T_C. Because the work W_{in} needed to operate the refrigerator is exactly the same as the useful work W_{out} done by the heat engine, we can use the heat engine, as shown in FIGURE 19.18b, to drive the refrigerator. The heat Q_C the engine exhausts to the cold reservoir is exactly the same as the heat Q_C the refrigerator extracts from the cold reservoir. Similarly, the heat Q_H the engine extracts from the hot reservoir matches the heat Q_H the refrigerator exhausts to the hot reservoir. Consequently, there is no net heat transfer in either direction. The refrigerator exactly replaces all the heat energy that had been transferred out of the hot reservoir by the heat engine.

You may want to compare the reasoning used here with the reasoning we used with Figure 19.11. There we tried to use the output of a "perfect" heat engine to run a refrigerator but did *not* succeed.

A Perfectly Reversible Engine Has Maximum Efficiency

Now we've arrived at the critical step in the reasoning. Suppose I claim to have a heat engine that can operate between temperatures T_H and T_C with *more* efficiency than a perfectly reversible engine. FIGURE 19.19 shows the output of this heat engine operating the same perfectly reversible refrigerator that we used in Figure 19.18b.

FIGURE 19.19 A heat engine more efficient than a perfectly reversible engine could be used to violate the second law of thermodynamics.

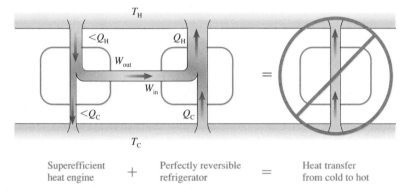

| Superefficient heat engine | $+$ | Perfectly reversible refrigerator | $=$ | Heat transfer from cold to hot |

Recall that the thermal efficiency and the work of a heat engine are

$$\eta = \frac{W_{out}}{Q_H} \quad \text{and} \quad W_{out} = Q_H - Q_C$$

If the new heat engine is more efficient than the perfectly reversible engine it replaces, it needs *less* heat Q_H from the hot reservoir to perform the *same* work W_{out}. If Q_H is less while W_{out} is the same, then Q_C must also be less. That is, the new heat engine exhausts less heat to the cold reservoir than does the perfectly reversible heat engine.

When this new heat engine drives the perfectly reversible refrigerator, the heat it exhausts to the cold reservoir is *less* than the heat extracted from the cold reservoir by the refrigerator. Similarly, this engine extracts *less* heat from the hot reservoir than the refrigerator exhausts. Thus the net result of using this superefficient heat engine to operate a perfectly reversible refrigerator is that heat is transferred from the cold reservoir to the hot reservoir *without outside assistance.*

But this can't happen. It would violate the second law of thermodynamics. Hence we have to conclude that no heat engine operating between reservoirs at temperatures T_H and T_C can be more efficient than a perfectly reversible engine. This very important conclusion is another version of the second law:

> **Second law, informal statement #5** No heat engine operating between reservoirs at temperatures T_H and T_C can be more efficient than a perfectly reversible engine operating between these temperatures.

The answer to our question "Is there a maximum η that cannot be exceeded?" is a clear Yes! The maximum possible efficiency η_{max} is that of a perfectly reversible engine. Because the perfectly reversible engine is an idealization, any real engine will have an efficiency less than η_{max}.

A similar argument shows that no refrigerator can be more efficient than a perfectly reversible refrigerator. If we had such a refrigerator, and if we ran it with the output of a perfectly reversible heat engine, we could transfer heat from cold to hot with no outside assistance. Thus:

> **Second law, informal statement #6** No refrigerator operating between reservoirs at temperatures T_H and T_C can have a coefficient of performance larger than that of a perfectly reversible refrigerator operating between these temperatures.

Conditions for a Perfectly Reversible Engine

This argument tells us that η_{max} and K_{max} exist, but it doesn't tell us what they are. Our final task will be to "design" and analyze a perfectly reversible engine. Under what conditions is an engine reversible?

An engine transfers energy by both mechanical and thermal interactions. Mechanical interactions are pushes and pulls. The environment does work on the system, transferring energy into the system by pushing in on a piston. The system transfers energy back to the environment by pushing out on the piston.

The energy transferred by a moving piston is perfectly reversible, returning the system to its initial state, with no change of temperature or pressure, only if the motion is *frictionless*. The slightest bit of friction will prevent the mechanical transfer of energy from being perfectly reversible.

The circumstances under which heat transfer can be *completely* reversed aren't quite so obvious. After all, Chapter 18 emphasized the *irreversible* nature of heat transfer. If objects A and B are in thermal contact, with $T_A > T_B$, then heat energy is transferred from A to B. But the second law of thermodynamics prohibits a heat transfer from B back to A. Heat transfer through a temperature *difference* is an irreversible process.

But suppose $T_A = T_B$. With no temperature difference, any heat that is transferred from A to B can, at a later time, be transferred from B back to A. This transfer wouldn't violate the second law, which prohibits only heat transfer from a colder object to a hotter object. Now you might object, and rightly so, that heat *can't* move from A to B if they are at the same temperature because heat, by definition, is the energy transferred between two objects at different temperatures.

This is true, so let's consider a limiting case in which $T_A = T_B + dT$. The temperature difference is infinitesimal. Heat is transferred from A to B, but *very slowly!* If you later try to make the heat move from B back to A, the second law will prevent you from doing so with perfect precision. But because the temperature difference is infinitesimal, you'll be missing only an infinitesimal amount dQ of heat. You can transfer heat reversibly in the limit $dT \rightarrow 0$, but you must be prepared to spend an infinite amount of time doing so.

Thus the thermal transfer of energy is reversible if the heat is transferred infinitely slowly in an isothermal process. This is an idealization, but so are completely frictionless processes. Nonetheless, we can now say that a perfectly reversible engine must use only two types of processes:

1. Frictionless mechanical interactions with no heat transfer ($Q = 0$), and
2. Thermal interactions in which heat is transferred in an isothermal process ($\Delta E_{th} = 0$).

Any engine that uses only these two types of processes is called a **Carnot engine**. A Carnot engine is a perfectly reversible engine; thus it has the maximum possible thermal efficiency η_{max} and, if operated as a refrigerator, the maximum possible coefficient of performance K_{max}.

19.6 The Carnot Cycle

The definition of a Carnot engine does not specify whether the engine's working substance is a gas or a liquid. It makes no difference. Our argument that a perfectly reversible engine is the most efficient possible heat engine depended only on the engine's reversibility. Consequently, **any Carnot engine operating between T_H and T_C must have exactly the same efficiency as any other Carnot engine operating between the same two energy reservoirs.** If we can determine the thermal efficiency of one Carnot engine, we'll know the efficiency of all Carnot engines. Because liquids and phase changes are complicated, we'll analyze a Carnot engine that uses an ideal gas.

Designing a Carnot Engine

The **Carnot cycle** is an ideal-gas cycle that consists of the two adiabatic processes ($Q = 0$) and two isothermal processes ($\Delta E_{th} = 0$) shown in FIGURE 19.20. These are the two types of processes allowed in a perfectly reversible gas engine. As a Carnot cycle operates,

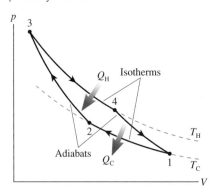

FIGURE 19.20 The Carnot cycle is perfectly reversible.

1. The gas is isothermally compressed while in thermal contact with the cold reservoir at temperature T_C. Heat energy $Q_C = |Q_{12}|$ is removed from the gas as it is compressed in order to keep the temperature constant. The compression must take place extremely slowly because there can be only an infinitesimal temperature difference between the gas and the reservoir.
2. The gas is adiabatically compressed while thermally isolated from the environment. This compression increases the gas temperature until it matches temperature T_H of the hot reservoir. No heat is transferred during this process.
3. After reaching maximum compression, the gas expands isothermally at temperature T_H. Heat $Q_H = Q_{34}$ is transferred from the hot reservoir into the gas as it expands in order to keep the temperature constant.
4. Finally, the gas expands adiabatically, with $Q = 0$, until the temperature decreases back to T_C.

Work is done in all four processes of the Carnot cycle, but heat is transferred only during the two isothermal processes.

The thermal efficiency of any heat engine is

$$\eta = \frac{W_{out}}{Q_H} = 1 - \frac{Q_C}{Q_H}$$

We can determine η_{Carnot} by finding the heat transfer in the two isothermal processes.

Process $1 \rightarrow 2$: Table 19.1 gives us the heat transfer in an isothermal process at temperature T_C:

$$Q_{12} = (W_s)_{12} = nRT_C \ln\left(\frac{V_2}{V_1}\right) = -nRT_C \ln\left(\frac{V_1}{V_2}\right) \qquad (19.20)$$

$V_1 > V_2$, so the logarithm on the right is positive. Q_{12} is negative because heat is transferred out of the system, but Q_C is simply the *amount* of heat transferred to the cold reservoir:

$$Q_C = |Q_{12}| = nRT_C \ln\left(\frac{V_1}{V_2}\right) \qquad (19.21)$$

Process $3 \rightarrow 4$: Similarly, the heat transferred in the isothermal expansion at temperature T_H is

$$Q_H = Q_{34} = (W_s)_{34} = nRT_H \ln\left(\frac{V_4}{V_3}\right) \qquad (19.22)$$

Thus the thermal efficiency of the Carnot cycle is

$$\eta_{Carnot} = 1 - \frac{Q_C}{Q_H} = 1 - \frac{T_C \ln(V_1/V_2)}{T_H \ln(V_4/V_3)} \qquad (19.23)$$

We can simplify this expression. During the two adiabatic processes,

$$T_C V_2^{\gamma-1} = T_H V_3^{\gamma-1} \qquad \text{and} \qquad T_C V_1^{\gamma-1} = T_H V_4^{\gamma-1} \qquad (19.24)$$

An algebraic rearrangement gives

$$V_2 = V_3 \left(\frac{T_H}{T_C}\right)^{1/(\gamma-1)} \qquad \text{and} \qquad V_1 = V_4 \left(\frac{T_H}{T_C}\right)^{1/(\gamma-1)} \qquad (19.25)$$

from which it follows that

$$\frac{V_1}{V_2} = \frac{V_4}{V_3} \qquad (19.26)$$

Consequently, the two logarithms in Equation 19.23 cancel and we're left with the result that the thermal efficiency of a Carnot engine operating between a hot reservoir at temperature T_H and a cold reservoir at temperature T_C is

$$\eta_{\text{Carnot}} = 1 - \frac{T_C}{T_H} \qquad \text{(Carnot thermal efficiency)} \qquad (19.27)$$

This remarkably simple result, an efficiency that depends only on the ratio of the temperatures of the hot and cold reservoirs, is Carnot's legacy to thermodynamics.

NOTE ▶ Temperatures T_H and T_C are *absolute* temperatures. ◀

EXAMPLE 19.4 **A Carnot engine**

A Carnot engine is cooled by water at $T_C = 10°C$. What temperature must be maintained in the hot reservoir of the engine to have a thermal efficiency of 70%?

MODEL The efficiency of a Carnot engine depends only on the temperatures of the hot and cold reservoirs.

SOLVE The thermal efficiency $\eta_{\text{Carnot}} = 1 - T_C/T_H$ can be rearranged to give

$$T_H = \frac{T_C}{1 - \eta_{\text{Carnot}}} = 943 \text{ K} = 670°C$$

where we used $T_C = 283$ K.

ASSESS A "real" engine would need a higher temperature than this to provide 70% efficiency because no real engine will match the Carnot efficiency.

EXAMPLE 19.5 **A real engine**

The heat engine of Example 19.2 had a highest temperature of 2700 K, a lowest temperature of 300 K, and a thermal efficiency of 15%. What is the efficiency of a Carnot engine operating between these two temperatures?

SOLVE The Carnot efficiency is

$$\eta_{\text{Carnot}} = 1 - \frac{T_C}{T_H} = 1 - \frac{300 \text{ K}}{2700 \text{ K}} = 0.89 = 89\%$$

ASSESS The thermodynamic cycle used in Example 19.2 doesn't come anywhere close to the Carnot efficiency.

The Maximum Efficiency

In Section 19.2 we tried to invent a perfect engine with $\eta = 1$ and $Q_C = 0$. We found that we could not do so without violating the second law, so no engine can have $\eta = 1$. However, that example didn't rule out an engine with $\eta = 0.9999$. Further analysis has now shown that no heat engine operating between energy reservoirs at temperatures T_H and T_C can be more efficient than a perfectly reversible engine operating between these temperatures.

We've now reached the endpoint of this line of reasoning by establishing an exact result for the thermal efficiency of a perfectly reversible engine, the Carnot engine. We can summarize our conclusions:

> **Second law, informal statement #7** No heat engine operating between energy reservoirs at temperatures T_H and T_C can exceed the Carnot efficiency
>
> $$\eta_{\text{Carnot}} = 1 - \frac{T_C}{T_H}$$

As Example 19.5 showed, real engines usually fall well short of the Carnot limit.

We also found that no refrigerator can exceed the coefficient of performance of a perfectly reversible refrigerator. We'll leave the proof as a homework problem, but an analysis very similar to that above shows that the coefficient of performance of a Carnot refrigerator is

$$K_{\text{Carnot}} = \frac{T_C}{T_H - T_C} \qquad \text{(Carnot coefficient of performance)} \qquad (19.28)$$

Thus we can state:

> **Second law, informal statement #8** No refrigerator operating between energy reservoirs at temperatures T_H and T_C can exceed the Carnot coefficient of performance
>
> $$K_{Carnot} = \frac{T_C}{T_H - T_C}$$

EXAMPLE 19.6   Brayton versus Carnot

The Brayton-cycle refrigerator of Example 19.3 had coefficient of performance $K = 1.1$. Compare this to the limit set by the second law of thermodynamics.

SOLVE Example 19.3 found that the reservoir temperatures had to be $T_C \geq 250$ K and $T_H \leq 381$ K. A Carnot refrigerator operating between 250 K and 381 K has

$$K_{Carnot} = \frac{T_C}{T_H - T_C} = \frac{250 \text{ K}}{381 \text{ K} - 250 \text{ K}} = 1.9$$

ASSESS This is the minimum value of K_{Carnot}. It will be even higher if $T_C > 250$ K or $T_H < 381$ K. The coefficient of performance of the reasonably realistic refrigerator of Example 19.3 is less than 60% of the limiting value.

Statements #7 and #8 of the second law are a major result of this chapter, one with profound implications. The efficiency limit of a heat engine is set by the temperatures of the hot and cold reservoirs. High efficiency requires $T_C/T_H \ll 1$ and thus $T_H \gg T_C$. However, practical realities often prevent T_H from being significantly larger than T_C, in which case the engine cannot possibly have a large efficiency. This limit on the efficiency of heat engines is a consequence of the second law of thermodynamics.

EXAMPLE 19.7   Generating electricity

An electric power plant boils water to produce high-pressure steam at 400°C. The high-pressure steam spins a turbine as it expands, then the turbine spins the generator. The steam is then condensed back to water in an ocean-cooled heat exchanger at 25°C. What is the *maximum* possible efficiency with which heat energy can be converted to electric energy?

MODEL The maximum possible efficiency is that of a Carnot engine operating between these temperatures.

SOLVE The Carnot efficiency depends on absolute temperatures, so we must use $T_H = 400°\text{C} = 673$ K and $T_C = 25°\text{C} = 298$ K. Then

$$\eta_{max} = 1 - \frac{298}{673} = 0.56 = 56\%$$

ASSESS This is an upper limit. Real coal-, oil-, gas-, and nuclear-heated steam generators actually operate at $\approx 35\%$ thermal efficiency, converting only about one-third of the fuel energy to electric energy while exhausting about two-thirds of the energy to the environment as waste heat. (The heat *source* has nothing to do with the efficiency. All it does is boil water.) Not much can be done to alter the low-temperature limit. The high-temperature limit is determined by the maximum temperature and pressure the boiler and turbine can withstand. The efficiency of electricity generation is far less than most people imagine, but it is an unavoidable consequence of the second law of thermodynamics.

A limit on the efficiency of heat engines was not expected. We are used to thinking in terms of energy conservation, so it comes as no surprise that we cannot make an engine with $\eta > 1$. But the limits arising from the second law were not anticipated, nor are they obvious. Nonetheless, they are a very real fact of life and a very real constraint on any practical device. No one has ever invented a machine that exceeds the second-law limits, and we have seen that the maximum efficiency for realistic engines is surprisingly low.

STOP TO THINK 19.5 Could this heat engine be built?

a. Yes.
b. No.
c. It's impossible to tell without knowing what kind of cycle it uses.

CHALLENGE EXAMPLE 19.8 **Calculating efficiency**

A heat engine using a monatomic ideal gas goes through the following closed cycle:

- Isochoric heating until the pressure is doubled.
- Isothermal expansion until the pressure is restored to its initial value.
- Isobaric compression until the volume is restored to its initial value.

What is the thermal efficiency of this heat engine? What would be the thermal efficiency of a Carnot engine operating between the highest and lowest temperatures reached by this engine?

MODEL The cycle consists of three familiar processes; we'll need to analyze each. The amount of work and heat will depend on the quantity of gas, which we don't know, but efficiency is a work-to-heat ratio that is independent of the amount of gas.

VISUALIZE FIGURE 19.21 shows the cycle. The initial pressure, volume, and temperature are p, V, and T. The isochoric process increases the pressure to $2p$ and, because the ratio p/T is constant in an isochoric process, increases the temperature to $2T$. The isothermal expansion is along the $2T$ isotherm. The product pV is constant in an isothermal process, so the volume doubles to $2V$ as the pressure returns to p.

FIGURE 19.21 The pV cycle of the heat engine.

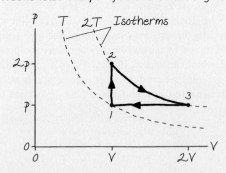

SOLVE We know, symbolically, the state variables at each corner of the pV diagram. That is sufficient for calculating W_s, Q, and ΔE_{th}.

Process $1 \rightarrow 2$: An isochoric process has $W_s = 0$ and

$$Q = \Delta E_{th} = nC_V \Delta T = \tfrac{3}{2} nRT$$

where we used $C_V = \tfrac{3}{2}R$ for a monatomic gas and $\Delta T = 2T - T = T$.

Process $2 \rightarrow 3$: An isothermal process has $\Delta E_{th} = 0$ and

$$Q = W_s = nR(2T)\ln(2V/V) = (2\ln 2)nRT$$

Here we used the Table 19.1 result for the work done in an isothermal process.

Process $3 \rightarrow 1$: The work done by the gas is the area under the curve, which is negative because $\Delta V = V - 2V = -V$ in the compression:

$$W_s = \text{area} = p\,\Delta V = -pV = -nRT$$

We used the ideal-gas law in the last step to express the result in terms of n and T. The heat transfer is also negative because $\Delta T = T - 2T = -T$:

$$Q = nC_P \Delta T = -\tfrac{5}{2} nRT$$

where we used $C_P = \tfrac{5}{2}R$ for a monatomic gas. Based on the first law, $\Delta E_{th} = Q - W_s = -\tfrac{3}{2} nRT$.

Summing over the three processes, we see that $(\Delta E_{th})_{net} = 0$, as expected, and

$$W_{out} = (2\ln 2 - 1)nRT$$

Heat energy is supplied to the gas ($Q > 0$) in processes $1 \rightarrow 2$ and $2 \rightarrow 3$, so

$$Q_H = (2\ln 2 + \tfrac{3}{2})nRT$$

Thus the thermal efficiency of this heat engine is

$$\eta = \frac{W_{out}}{Q_H} = \frac{(2\ln 2 - 1)nRT}{(2\ln 2 + \tfrac{3}{2})nRT} = 0.134 = 13.4\%$$

A Carnot engine would be able to operate between a high temperature $T_H = 2T$ and a low temperature $T_C = T$. Its efficiency would be

$$\eta_{Carnot} = 1 - \frac{T_C}{T_H} = 1 - \frac{T}{2T} = 0.500 = 50.0\%$$

ASSESS As we anticipated, the thermal efficiency depends on the *shape* of the pV cycle but not on the quantity of gas or even on the values of p, V, or T. The heat engine's 13.4% efficiency is considerably less than the 50% maximum possible efficiency set by the second law of thermodynamics.

SUMMARY

The goal of Chapter 19 has been to study the physical principles that govern heat engines and refrigerators.

General Principles

Heat Engines

Devices that transform heat into work. They require two energy reservoirs at different temperatures.

Thermal efficiency

$$\eta = \frac{W_{out}}{Q_H} = \frac{\text{what you get}}{\text{what you pay}}$$

Second-law limit:

$$\eta \leq 1 - \frac{T_C}{T_H}$$

Refrigerators

Devices that use work to transfer heat from a colder object to a hotter object.

Coefficient of performance

$$K = \frac{Q_C}{W_{in}} = \frac{\text{what you get}}{\text{what you pay}}$$

Second-law limit:

$$K \leq \frac{T_C}{T_H - T_C}$$

Important Concepts

A perfectly reversible engine (a **Carnot engine**) can be operated as either a heat engine or a refrigerator between the same two energy reservoirs by reversing the cycle and with no other changes.

- A **Carnot heat engine** has the maximum possible thermal efficiency of any heat engine operating between T_H and T_C:

$$\eta_{Carnot} = 1 - \frac{T_C}{T_H}$$

- A **Carnot refrigerator** has the maximum possible coefficient of performance of any refrigerator operating between T_H and T_C:

$$K_{Carnot} = \frac{T_C}{T_H - T_C}$$

The Carnot cycle for a gas engine consists of two isothermal processes and two adiabatic processes.

An energy reservoir is a part of the environment so large in comparison to the system that its temperature doesn't change as the system extracts heat energy from or exhausts heat energy to the reservoir. All heat engines and refrigerators operate between two energy reservoirs at different temperatures T_H and T_C.

The **work** W_s done *by* the system has the opposite sign to the work done *on* the system.

$W_s =$ area under pV curve

Applications

To analyze a heat engine or refrigerator:

MODEL Identify each process in the cycle.

VISUALIZE Draw the pV diagram of the cycle.

SOLVE There are several steps:

- Determine p, V, and T at the beginning and end of each process.
- Calculate ΔE_{th}, W_s, and Q for each process.
- Determine W_{in} or W_{out}, Q_H, and Q_C.
- Calculate $\eta = W_{out}/Q_H$ or $K = Q_C/W_{in}$.

ASSESS Verify $(\Delta E_{th})_{net} = 0$. Check signs.

Terms and Notation

thermodynamics	closed-cycle device	coefficient of performance, K	Carnot engine
energy reservoir	thermal efficiency, η	heat exchanger	Carnot cycle
energy-transfer diagram	waste heat	pressure ratio, r_p	
heat engine	refrigerator	perfectly reversible engine	

CONCEPTUAL QUESTIONS

1. In going from i to f in each of the three processes of FIGURE Q19.1, is work done *by* the system ($W < 0$, $W_s > 0$), is work done *on* the system ($W > 0$, $W_s < 0$), or is *no* net work done?

FIGURE Q19.1

2. Rank in order, from largest to smallest, the amount of work $(W_s)_1$ to $(W_s)_4$ done by the gas in each of the cycles shown in FIGURE Q19.2. Explain.

FIGURE Q19.2

3. Could you have a heat engine with $\eta > 1$? Explain.

4. FIGURE Q19.4 shows the pV diagram of a heat engine. During which stage or stages is (a) heat added to the gas, (b) heat removed from the gas, (c) work done on the gas, and (d) work done by the gas?

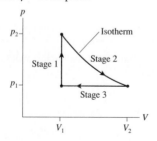

FIGURE Q19.4

5. Rank in order, from largest to smallest, the thermal efficiencies η_1 to η_4 of the four heat engines in FIGURE Q19.5. Explain.

FIGURE Q19.5

6. FIGURE Q19.6 shows the thermodynamic cycles of two heat engines. Which heat engine has the larger thermal efficiency? Or are they the same? Explain.

FIGURE Q19.6

7. A heat engine satisfies $W_{out} = Q_{net}$. Why is there no ΔE_{th} term in this relationship?

8. Do the energy-transfer diagrams in FIGURE Q19.8 represent possible heat engines? If not, what is wrong?

FIGURE Q19.8

9. Do the energy-transfer diagrams in FIGURE Q19.9 represent possible refrigerators? If not, what is wrong?

FIGURE Q19.9

10. It gets pretty hot in your apartment. In browsing the Internet, you find a company selling small "room air conditioners." You place the air conditioner on the floor, plug it in, and—the advertisement says—it will lower the room temperature up to 10°F. Should you order one? Explain.

11. The first and second laws of thermodynamics are sometimes stated as "You can't win" and "You can't even break even." Do these sayings accurately characterize the laws of thermodynamics as applied to heat engines? Why or why not?

EXERCISES AND PROBLEMS

Problems labeled ▦ integrate material from earlier chapters.

Exercises

Section 19.1 Turning Heat into Work

Section 19.2 Heat Engines and Refrigerators

1. | A heat engine with a thermal efficiency of 40% does 100 J of work per cycle. How much heat is (a) extracted from the hot reservoir and (b) exhausted to the cold reservoir per cycle?

2. ‖ A heat engine does 200 J of work per cycle while exhausting 400 J of waste heat. What is the engine's thermal efficiency?

3. | A heat engine extracts 55 kJ of heat from the hot reservoir each cycle and exhausts 40 kJ of heat. What are (a) the thermal efficiency and (b) the work done per cycle?

4. ‖ A refrigerator requires 200 J of work and exhausts 600 J of heat per cycle. What is the refrigerator's coefficient of performance?

5. | 50 J of work are done per cycle on a refrigerator with a coefficient of performance of 4.0. How much heat is (a) extracted from the cold reservoir and (b) exhausted to the hot reservoir per cycle?

6. ‖ The power output of a car engine running at 2400 rpm is 500 kW. How much (a) work is done and (b) heat is exhausted per cycle if the engine's thermal efficiency is 20%? Give your answers in kJ.

7. ‖ A 32%-efficient electric power plant produces 900 MW of electric power and discharges waste heat into 20°C ocean water. Suppose the waste heat could be used to heat homes during the winter instead of being discharged into the ocean. A typical American house requires an average 20 kW for heating. How many homes could be heated with the waste heat of this one power plant?

8. ‖ 1.0 L of 20°C water is placed in a refrigerator. The refrigerator's motor must supply an extra 8.0 W of power to chill the water to 5°C in 1.0 h. What is the refrigerator's coefficient of performance?

Section 19.3 Ideal-Gas Heat Engines

Section 19.4 Ideal-Gas Refrigerators

9. ‖ The cycle of FIGURE EX19.9 consists of four processes. Make a table with rows labeled A to D and columns labeled ΔE_{th}, W_s, and Q. Fill each box in the table with +, −, or 0 to indicate whether the quantity increases, decreases, or stays the same during that process.

FIGURE EX19.9

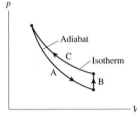

FIGURE EX19.10

10. ‖ The cycle of FIGURE EX19.10 consists of three processes. Make a table with rows labeled A–C and columns labeled ΔE_{th}, W_s, and Q. Fill each box in the table with +, −, or 0 to indicate whether the quantity increases, decreases, or stays the same during that process.

11. ‖ How much work is done per cycle by a gas following the pV trajectory of FIGURE EX19.11?

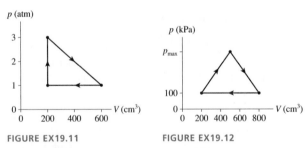

FIGURE EX19.11

FIGURE EX19.12

12. ‖ A gas following the pV trajectory of FIGURE EX19.12 does 60 J of work per cycle. What is p_{max}?

13. ‖ What are (a) W_{out} and Q_H and (b) the thermal efficiency for the heat engine shown in FIGURE EX19.13?

FIGURE EX19.13

FIGURE EX19.14

14. ‖ What are (a) W_{out} and Q_C and (b) the thermal efficiency for the heat engine shown in FIGURE EX19.14?

15. ‖ How much heat is exhausted to the cold reservoir by the heat engine shown in FIGURE EX19.15?

FIGURE EX19.15   FIGURE EX19.16

16. ‖ What are (a) the thermal efficiency and (b) the heat extracted from the hot reservoir for the heat engine shown in FIGURE EX19.16?

17. ‖ A heat engine uses a diatomic gas in a Brayton cycle. What is the engine's thermal efficiency if the gas volume is halved during the adiabatic compression?

18. ‖ What are (a) the heat extracted from the cold reservoir and (b) the coefficient of performance for the refrigerator shown in FIGURE EX19.18?

FIGURE EX19.18

Section 19.5 The Limits of Efficiency

Section 19.6 The Carnot Cycle

19. ‖ Which, if any, of the heat engines in FIGURE EX19.19 violate (a) the first law of thermodynamics or (b) the second law of thermodynamics? Explain.

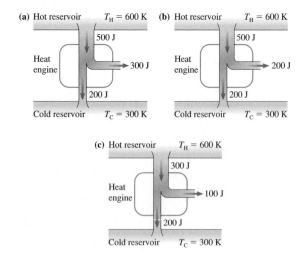

FIGURE EX19.19

20. ‖ Which, if any, of the refrigerators in FIGURE EX19.20 violate (a) the first law of thermodynamics or (b) the second law of thermodynamics? Explain.

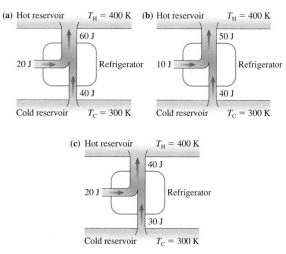

FIGURE EX19.20

21. ‖ At what cold-reservoir temperature (in °C) would a Carnot engine with a hot-reservoir temperature of 427°C have an efficiency of 60%?

22. ‖ A heat engine does 10 J of work and exhausts 15 J of waste heat during each cycle.
   a. What is the engine's thermal efficiency?
   b. If the cold-reservoir temperature is 20°C, what is the minimum possible temperature in °C of the hot reservoir?

23. ‖ a. A heat engine does 200 J of work per cycle while exhausting 600 J of heat to the cold reservoir. What is the engine's thermal efficiency?
   b. A Carnot engine with a hot-reservoir temperature of 400°C has the same thermal efficiency. What is the cold-reservoir temperature in °C?

24. ‖ A Carnot engine operating between energy reservoirs at temperatures 300 K and 500 K produces a power output of 1000 W. What are (a) the thermal efficiency of this engine, (b) the rate of heat input, in W, and (c) the rate of heat output, in W?

25. ‖ A Carnot engine whose hot-reservoir temperature is 400°C has a thermal efficiency of 40%. By how many degrees should the temperature of the cold reservoir be decreased to raise the engine's efficiency to 60%?

26. ‖ A heat engine operating between energy reservoirs at 20°C and 600°C has 30% of the maximum possible efficiency. How much energy must this engine extract from the hot reservoir to do 1000 J of work?

27. ‖ A heat engine operating between a hot reservoir at 500°C and a cold reservoir at 0°C is 60% as efficient as a Carnot engine. If this heat engine and the Carnot engine do the same amount of work, what is the ratio $Q_H/(Q_H)_{Carnot}$?

28. ‖ A Carnot refrigerator operating between −20°C and +20°C extracts heat from the cold reservoir at the rate 200 J/s. What are (a) the coefficient of performance of this refrigerator, (b) the rate at which work is done on the refrigerator, and (c) the rate at which heat is exhausted to the hot side?

29. ‖ The coefficient of performance of a refrigerator is 5.0. The compressor uses 10 J of energy per cycle.
    a. How much heat energy is exhausted per cycle?
    b. If the hot-reservoir temperature is 27°C, what is the lowest possible temperature in °C of the cold reservoir?

30. ‖ A Carnot heat engine with thermal efficiency $\frac{1}{3}$ is run backward as a Carnot refrigerator. What is the refrigerator's coefficient of performance?

Problems

31. ‖ The engine that powers a crane burns fuel at a flame temperature of 2000°C. It is cooled by 20°C air. The crane lifts a 2000 kg steel girder 30 m upward. How much heat energy is transferred to the engine by burning fuel if the engine is 40% as efficient as a Carnot engine?

32. ‖‖ 100 mL of water at 15°C is placed in the freezer compartment of a refrigerator with a coefficient of performance of 4.0. How much heat energy is exhausted into the room as the water is changed to ice at 15°C?

33. ‖ Prove that the work done in an adiabatic process i → f is $W_s = (p_f V_f - p_i V_i)/(1 - \gamma)$.

34. ‖ A Carnot refrigerator operates between reservoirs at −20°C and 50°C in a 25°C room. The refrigerator is a 40 cm × 40 cm × 40 cm box. Five of the walls are perfect insulators, but the sixth is a 1.0-cm-thick piece of stainless steel. What electric power does the refrigerator require to maintain the inside temperature at −20°C?

35. ‖ Prove that the coefficient of performance of a Carnot refrigerator is $K_{Carnot} = T_C/(T_H - T_C)$.

36. ‖ An ideal refrigerator utilizes a Carnot cycle operating between 0°C and 25°C. To turn 10 kg of liquid water at 0°C into 10 kg of ice at 0°C, (a) how much heat is exhausted into the room and (b) how much energy must be supplied to the refrigerator?

37. ‖ There has long been an interest in using the vast quantities of thermal energy in the oceans to run heat engines. A heat engine needs a temperature *difference*, a hot side and a cold side. Conveniently, the ocean surface waters are warmer than the deep ocean waters. Suppose you build a floating power plant in the tropics where the surface water temperature is ≈ 30°C. This would be the hot reservoir of the engine. For the cold reservoir, water would be pumped up from the ocean bottom where it is always ≈ 5°C. What is the maximum possible efficiency of such a power plant?

38. ‖ The ideal gas in a Carnot engine extracts 1000 J of heat energy during the isothermal expansion at 300°C. How much heat energy is exhausted during the isothermal compression at 50°C?

39. | The hot-reservoir temperature of a Carnot engine with 25% efficiency is 80°C higher than the cold-reservoir temperature. What are the reservoir temperatures, in °C?

40. ‖ A Carnot heat engine operates between reservoirs at 182°C and 0°C. If the engine extracts 25 J of energy from the hot reservoir per cycle, how many cycles will it take to lift a 10 kg mass a height of 10 m?

41. ‖ A Carnot refrigerator operates between reservoirs at 55°C and −20°C. If the engine exhausts 250 J of energy to the hot reservoir per cycle, how many cycles will it take to cool a 500 mL soda from 25°C to 5°C?

42. ‖ **FIGURE P19.42** shows a Carnot heat engine driving a Carnot refrigerator.

a. Determine Q_1, Q_2, Q_3, and Q_4.
b. Is Q_3 greater than, less than, or equal to Q_1?
c. Do these two devices, when operated together in this way, violate the second law?

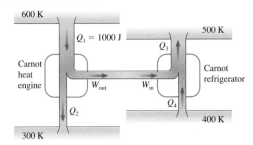

FIGURE P19.42

43. ‖ A Carnot heat engine and an ordinary refrigerator with coefficient of performance 2.00 operate between reservoirs at 350 K and 250 K. The work done by the Carnot heat engine drives the refrigerator. If the heat engine extracts 10.0 J of energy from the hot reservoir, how much energy does the refrigerator exhaust to the hot reservoir?

44. ‖ A heat engine running backward is called a refrigerator if its purpose is to extract heat from a cold reservoir. The same engine running backward is called a *heat pump* if its purpose is to exhaust warm air into the hot reservoir. Heat pumps are widely used for home heating. You can think of a heat pump as a refrigerator that is cooling the already cold outdoors and, with its exhaust heat Q_H, warming the indoors. Perhaps this seems a little silly, but consider the following. Electricity can be directly used to heat a home by passing an electric current through a heating coil. This is a direct, 100% conversion of work to heat. That is, 15 kW of electric power (generated by doing work at the rate of 15 kJ/s at the power plant) produces heat energy inside the home at a rate of 15 kJ/s. Suppose that the neighbor's home has a heat pump with a coefficient of performance of 5.0, a realistic value.
    a. How much electric power (in kW) does the heat pump use to deliver 15 kJ/s of heat energy to the house?
    b. An average price for electricity is about 40 MJ per dollar. A furnace or heat pump will run typically 200 hours per month during the winter. What does one month's heating cost in the home with a 15 kW electric heater and in the home of the neighbor who uses an equivalent heat pump?

45. ‖ You and your roommates need a new refrigerator. At the appliance store, the salesman shows you the DreamFridge. According to its sticker, the DreamFridge uses a mere 100 W of power to remove 100 kJ of heat per minute from the 2°C interior. According to the fine print on the sticker, this claim is true in a 22°C kitchen. Should you buy? Explain.

46. ‖ Three engineering students submit their solutions to a design problem in which they were asked to design an engine that operates between temperatures 300 K and 500 K. The heat input/output and work done by their designs are shown in the following table:

Student	Q_H	Q_C	W_{out}
1	250 J	140 J	110 J
2	250 J	170 J	90 J
3	250 J	160 J	90 J

Critique their designs. Are they acceptable or not? Is one better than the others? Explain.

47. ‖ A typical coal-fired power plant burns 300 metric tons of coal *every hour* to generate 750 MW of electricity. 1 metric ton = 1000 kg. The density of coal is 1500 kg/m^3 and its heat of combustion is 28 MJ/kg. Assume that *all* heat is transferred from the fuel to the boiler and that *all* the work done in spinning the turbine is transformed into electric energy.
    a. Suppose the coal is piled up in a 10 m × 10 m room. How tall must the pile be to operate the plant for one day?
    b. What is the power plant's thermal efficiency?

48. ‖ A nuclear power plant generates 3000 MW of heat energy from nuclear reactions in the reactor's core. This energy is used to boil water and produce high-pressure steam at 300°C. The steam spins a turbine, which produces 1000 MW of electric power, then the steam is condensed and the water is cooled to 25°C before starting the cycle again.
    a. What is the maximum possible thermal efficiency of the power plant?
    b. What is the plant's actual efficiency?
    c. Cooling water from a river flows through the condenser (the low-temperature heat exchanger) at the rate of 1.2 × 10^8 L/h (≈ 30 million gallons per hour). If the river water enters the condenser at 18°C, what is its exit temperature?

49. ‖ The electric output of a power plant is 750 MW. Cooling water flows through the power plant at the rate 1.0 × 10^8 L/h. The cooling water enters the plant at 16°C and exits at 27°C. What is the power plant's thermal efficiency?

50. ‖ a. A large nuclear power plant has a power output of 1000 MW. In other words, it generates electric energy at the rate 1000 MJ/s. How much energy does this power plant supply in one day?
    b. The oceans are vast. How much energy could be extracted from 1 km^3 of water if its temperature were decreased by 1°C? For simplicity, assume fresh water.
    c. A friend of yours who is an inventor comes to you with an idea. He has done the calculations that you just did in parts a and b, and he's concluded that a few cubic kilometers of ocean water could meet most of the energy needs of the United States. This is an insignificant fraction of the U.S. coastal waters. In addition, the oceans are constantly being reheated by the sun, so energy obtained from the ocean is essentially solar energy. He has sketched out some design plans—highly secret, of course, because they're not patented—and now he needs some investors to provide money for a prototype. A working prototype will lead to a patent. As an initial investor, you'll receive a fraction of all future royalties. Time is of the essence because a rival inventor is working on the same idea. He needs $10,000 from you right away. You could make millions if it works out. Will you invest? If so, explain why. If not, why not? Either way, your explanation should be based on scientific principles. Sketches and diagrams are a reasonable part of an explanation.

51. ‖ An air conditioner removes 5.0 × 10^5 J/min of heat from a house and exhausts 8.0 × 10^5 J/min to the hot outdoors.
    a. How much power does the air conditioner's compressor require?
    b. What is the air conditioner's coefficient of performance?

52. ‖ A heat engine using 1.0 mol of a monatomic gas follows the cycle shown in FIGURE P19.52. 3750 J of heat energy is transferred to the gas during process 1 → 2.
    a. Determine W_s, Q, and ΔE_{th} for each of the four processes in this cycle. Display your results in a table.
    b. What is the thermal efficiency of this heat engine?

FIGURE P19.52

FIGURE P19.53

53. ‖ A heat engine using a diatomic gas follows the cycle shown in FIGURE P19.53. Its temperature at point 1 is 20°C.
    a. Determine W_s, Q, and ΔE_{th} for each of the three processes in this cycle. Display your results in a table.
    b. What is the thermal efficiency of this heat engine?
    c. What is the power output of the engine if it runs at 500 rpm?

54. ‖ FIGURE P19.54 shows the cycle for a heat engine that uses a gas having $\gamma = 1.25$. The initial temperature is $T_1 = 300$ K, and this engine operates at 20 cycles per second.
    a. What is the power output of the engine?
    b. What is the engine's thermal efficiency?

FIGURE P19.54

55. ‖ A heat engine using a monatomic gas follows the cycle shown in FIGURE P19.55.
    a. Find W_s, Q, and ΔE_{th} for each process in the cycle. Display your results in a table.
    b. What is the thermal efficiency of this heat engine?

FIGURE P19.55

56. ‖ A heat engine uses a diatomic gas that follows the pV cycle in FIGURE P19.56.
    a. Determine the pressure, volume, and temperature at point 2.
    b. Determine ΔE_{th}, W_s, and Q for each of the three processes. Put your results in a table for easy reading.
    c. How much work does this engine do per cycle and what is its thermal efficiency?

FIGURE P19.56

57. ‖ A heat engine uses a diatomic gas that follows the pV cycle in **FIGURE P19.57**.
   a. Determine the pressure, volume, and temperature at point 1.
   b. Determine ΔE_{th}, W_s, and Q for each of the three processes. Put your results in a table for easy reading.
   c. How much work does this engine do per cycle and what is its thermal efficiency?

FIGURE P19.57

58. ‖ A refrigerator using helium gas operates on the reversed cycle shown in **FIGURE P19.58**. What are the refrigerator's (a) coefficient of performance and (b) power input if it operates at 60 cycles per second?

FIGURE P19.58

59. ‖ A heat engine using 120 mg of helium as the working substance follows the cycle shown in **FIGURE P19.59**.
   a. Determine the pressure, temperature, and volume of the gas at points 1, 2, and 3.
   b. What is the engine's thermal efficiency?
   c. What is the maximum possible efficiency of a heat engine that operates between T_{max} and T_{min}?

FIGURE P19.59

60. ‖ The heat engine shown in **FIGURE P19.60** uses 2.0 mol of a monatomic gas as the working substance.
   a. Determine T_1, T_2, and T_3.
   b. Make a table that shows ΔE_{th}, W_s, and Q for each of the three processes.
   c. What is the engine's thermal efficiency?

FIGURE P19.60

61. ‖ The heat engine shown in **FIGURE P19.61** uses 0.020 mol of a diatomic gas as the working substance.
   a. Determine T_1, T_2, and T_3.
   b. Make a table that shows ΔE_{th}, W_s, and Q for each of the three processes.
   c. What is the engine's thermal efficiency?

FIGURE P19.61

62. ‖ A heat engine using a diatomic ideal gas goes through the following closed cycle:
   - Isothermal compression until the volume is halved.
   - Isobaric expansion until the volume is restored to its initial value.
   - Isochoric cooling until the pressure is restored to its initial value.

   What are the thermal efficiencies of (a) this heat engine and (b) a Carnot engine operating between the highest and lowest temperatures reached by this engine?

63. ‖ A heat engine with 0.20 mol of a monatomic ideal gas initially fills a 2000 cm³ cylinder at 600 K. The gas goes through the following closed cycle:
   - Isothermal expansion to 4000 cm³.
   - Isochoric cooling to 300 K.
   - Isothermal compression to 2000 cm³.
   - Isochoric heating to 600 K.

   How much work does this engine do per cycle and what is its thermal efficiency?

64. ‖ **FIGURE P19.64** is the pV diagram of Example 19.2, but now the device is operated in reverse.
   a. During which processes is heat transferred into the gas?
   b. Is this Q_H, heat extracted from a hot reservoir, or Q_C, heat extracted from a cold reservoir? Explain.
   c. Determine the values of Q_H and Q_C.
   Hint: The calculations have been done in Example 19.2 and do not need to be repeated. Instead, you need to determine which processes now contribute to Q_H and which to Q_C.
   d. Is the area inside the curve W_{in} or W_{out}? What is its value?
   e. The device is now being operated in a ccw cycle. Is it a refrigerator? Explain.

FIGURE P19.64

In Problems 65 through 68 you are given the equation(s) used to solve a problem. For each of these, you are to

a. Write a realistic problem for which this is the correct equation(s).
b. Finish the solution of the problem.

65. $0.80 = 1 - (0°C + 273)/(T_H + 273)$

66. $4.0 = Q_C/W_{in}$

    $Q_H = 100$ J

67. $0.20 = 1 - Q_C/Q_H$

    $W_{out} = Q_H - Q_C = 20$ J

68. 400 kJ $= \frac{1}{2}(p_{max} - 100$ kPa$)(3.0$ m$^3 - 1.0$ m$^3)$

Challenge Problems

69. FIGURE CP19.69 shows a heat engine going through one cycle. The gas is diatomic. The masses are such that when the pin is removed, in steps 3 and 6, the piston does not move.
    a. Draw the pV diagram for this heat engine.
    b. How much work is done per cycle?
    c. What is this engine's thermal efficiency?

1. Start.   2. Heat to 3 atm.   3. Remove pin. Continue heating to 100 cm^3.   4. Insert pin. Remove mass.

5. Cool to 1.0 atm.   6. Remove pin. Continue cooling to 50 cm^3.   7. Insert pin. Add mass. Start again.

FIGURE CP19.69

70. FIGURE CP19.70 shows two insulated compartments separated by a thin wall. The left side contains 0.060 mol of helium at an initial temperature of 600 K and the right side contains 0.030 mol of helium at an initial temperature of 300 K. The compartment on the right is attached to a vertical cylinder, above which the air pressure is 1.0 atm. A 10-cm-diameter, 2.0 kg piston can slide without friction up and down the cylinder. Neither the cylinder diameter nor the volumes of the compartments are known.

a. What is the final temperature?
b. How much heat is transferred from the left side to the right side?
c. How high is the piston lifted due to this heat transfer?
d. What fraction of the heat is converted into work?

FIGURE CP19.70

71. The gasoline engine in your car can be modeled as the Otto cycle shown in FIGURE CP19.71. A fuel-air mixture is sprayed into the cylinder at point 1, where the piston is at its farthest distance from the spark plug. This mixture is compressed as the piston moves toward the spark plug during the adiabatic *compression stroke*. The spark plug fires at point 2, releasing heat energy that had been stored in the gasoline. The fuel burns so quickly that the piston doesn't have time to move, so the heating is an isochoric process. The hot, high-pressure gas then pushes the piston outward during the *power stroke*. Finally, an exhaust value opens to allow the gas temperature and pressure to drop back to their initial values before starting the cycle over again.

a. Analyze the Otto cycle and show that the work done per cycle is

$$W_{out} = \frac{nR}{1-\gamma}(T_2 - T_1 + T_4 - T_3)$$

b. Use the adiabatic connection between T_1 and T_2 and also between T_3 and T_4 to show that the thermal efficiency of the Otto cycle is

$$\eta = 1 - \frac{1}{r^{(\gamma-1)}}$$

where $r = V_{max}/V_{min}$ is the engine's *compression ratio*.

c. Graph η versus r out to $r = 30$ for a diatomic gas.

FIGURE CP19.71

72. FIGURE CP19.72 shows the Diesel cycle. It is similar to the Otto cycle (see Problem CP19.71), but there are two important differences. First, the fuel is not admitted until the air is fully compressed at point 2. Because of the high temperature at the end of an adiabatic compression, the fuel begins to burn spontaneously. (There are no spark plugs in a diesel engine!) Second,

combustion takes place more slowly, with fuel continuing to be injected. This makes the ignition stage a constant-pressure process. The cycle shown, for one cylinder of a diesel engine, has a *displacement* $V_{max} - V_{min}$ of 1000 cm³ and a compression ratio $r = V_{max}/V_{min} = 21$. These are typical values for a diesel truck. The engine operates with intake air ($\gamma = 1.40$) at 25°C and 1.0 atm pressure. The quantity of fuel injected into the cylinder has a heat of combustion of 1000 J.

a. Find p, V, and T at each of the four corners of the cycle. Display your results in a table.
b. What is the net work done by the cylinder during one full cycle?
c. What is the thermal efficiency of this engine?

d. What is the power output in kW and horsepower (1 hp = 746 W) of an eight-cylinder diesel engine running at 2400 rpm?

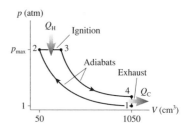

FIGURE CP19.72

STOP TO THINK ANSWERS

Stop to Think 19.1: $W_d > W_a = W_b > W_c$. $W_{out} = Q_H - Q_C$.

Stop to Think 19.2: b. Energy conservation requires $Q_H = Q_C + W_{in}$. The refrigerator will exhaust more heat out the back than it removes from the front. A refrigerator with an open door will heat the room rather than cool it.

Stop to Think 19.3: c. W_{out} = area inside triangle = 1000 J. $\eta = W_{out}/Q_H = (1000 \text{ J})/(4000 \text{ J}) = 0.25$.

Stop to Think 19.4: To conserve energy, the heat Q_H exhausted to the hot reservoir needs to be $Q_H = Q_C + W_{in} = 40 \text{ J} + 10 \text{ J} = 50 \text{ J}$, not 30 J. The numbers shown here would be appropriate to a heat engine if the energy-transfer arrows were all reversed.

Stop to Think 19.5: b. The efficiency of this engine would be $\eta = W_{out}/Q_H = 0.6$. That exceeds the Carnot efficiency $\eta_{Carnot} = 1 - T_C/T_H = 0.5$, so it is not possible.

Part IV had two important goals: first, to learn how energy is transformed; second, to establish a micro/macro connection in which we can understand the macroscopic properties of solids, liquids, and gases in terms of the microscopic motions of atoms and molecules. We have been quite successful. You have learned that:

■ Temperature is a measure of the thermal energy of the molecules in a system, and the average energy per molecule is simply $\frac{1}{2}k_B T$ per degree of freedom.

■ The pressure of a gas is due to collisions of the molecules with the walls of the container.

■ Heat is the energy transferred between two systems that have different temperatures. An important mechanism of heat transfer is molecular collisions at the boundary between the two systems.

■ Work, heat, and thermal energy can be transformed into each other in accord with the first law of thermodynamics, $\Delta E_{th} = W + Q$. This is a statement that energy is conserved.

■ Practical devices for turning heat into work, called heat engines, are limited in their efficiency by the second law of thermodynamics.

The knowledge structure of thermodynamics below summarizes the basic laws, diagramming our energy model and presenting our model of a heat engine in pictorial form. Thermodynamics, more than most topics in physics, can seem very "equation oriented." It's undeniable that there are more equations than we used in earlier parts of this text and more things to remember. But focusing on the equations is seeing only the trees, not the forest. A better strategy is to focus on the ideas embedded in the knowledge structure. You can find the necessary equations if you know how the ideas are connected, but memorizing all the equations won't help if you don't know which are relevant to different situations.

KNOWLEDGE STRUCTURE IV **Thermodynamics**

ESSENTIAL CONCEPTS	Work, heat, and thermal energy
BASIC GOALS	How is energy converted from one form to another?
	How are macroscopic properties related to microscopic behavior?

GENERAL PRINCIPLES	**First law of thermodynamics**	Energy is conserved, $\Delta E_{th} = W + Q$.
	Second law of thermodynamics	Heat is not spontaneously transferred from a colder object to a hotter object.

GAS LAWS AND PROCESSES Ideal-gas law $pV = nRT = Nk_B T$

- Isochoric process V = constant and $W = 0$
- Isothermal process T = constant and $\Delta E_{th} = 0$
- Isobaric process p = constant
- Adiabatic process $Q = 0$

Energy Transformation

Work on system   **Environment**   Work by system

$W > 0$ **System** $W < 0$
Thermal energy E_{th}
+
Energy in   Other state variables   Energy out
$p, V, T, n, M, \ldots$

$Q > 0$ First law: $\Delta E_{th} = W + Q$ $Q < 0$

Heat to system   Heat out of system

Work

Requires volume change

Gas: $W = -\int p\,dV$

$= -(\text{area under } pV \text{ curve})$

Thermal Energy

$E_{th} = \frac{1}{2}Nk_B T$ per degree of freedom

Heat

Requires temperature difference

$Q = Mc\Delta T$ or $nC\Delta T$

$Q = \pm ML$ for phase changes

Heat Engines

W_{out} = area inside pV curve

$= Q_H - Q_C$

$\eta = \dfrac{W_{out}}{Q_H}$

$\eta_{max} = \eta_{Carnot} = 1 - \dfrac{T_C}{T_H}$

Hot reservoir T_H
Q_H
Heat engine W_{out}
Q_C
Cold reservoir T_C

ONE STEP BEYOND

Order Out of Chaos

The second law predicts that systems will run down, that order will evolve toward disorder and randomness, and that complexity will give way to simplicity. But just look around you!

- Plants grow from simple seeds to complex entities.
- Single-cell fertilized eggs grow into complex adult organisms.
- Electric current passing through a "soup" of simple random molecules produces such complex chemicals as amino acids.
- Over the last billion or so years life has evolved from simple unicellular organisms to very complex forms.
- Knowledge and information seem to grow every year, not to fade away.

Everywhere we look, it seems, the second law is being violated. How can this be?

There is an important qualification in the second law of thermodynamics: It applies only to *isolated* systems, systems that do not exchange energy with their environment. The situation is entirely different if energy is transferred into or out of the system, and we cannot predict what will happen to the entropy of a nonisolated system. The popular-science literature is full of arguments and predictions that make incorrect use of the second law by trying to apply it to systems that are not isolated.

Systems that become *more* ordered as time passes, and in which the entropy decreases, are called *self-organizing systems.* All the examples listed above are self-organizing systems. One of the major characteristics of self-organizing systems is a substantial flow of energy *through* the system. For example, plants and animals take in energy from the sun or chemical energy from food, make use of that energy, and then give waste heat back to the environment via evaporation, decay, and other means. It is this energy flow that allows the systems to maintain, or even increase, a high degree of order and a very low entropy.

But—and this is the important point—the entropy of the *entire* system, including the earth and the sun, undergoes a significant *increase* so as to let selected subsystems decrease their entropy and become more ordered. The second law is not violated at all, but you must apply the second law to the combined systems that are interacting and not just to a single subsystem.

The snowflake in the photo is a beautiful example. As water freezes, the random motion of water molecules is transformed into a highly ordered crystal. The entropy of

A snowflake is a highly ordered arrangement of water molecules. The creation of a snowflake decreases the entropy of the water, but the second law of thermodynamics is not violated because the water molecules are not an isolated system.

the water molecules certainly decreases, but water doesn't freeze as an isolated system. For it to freeze, heat energy must be transferred from the water to the surrounding air. The entropy of the air increases by *more* than the entropy of the water decreases. Thus the *total* entropy of the water + air system increases when a snowflake is formed, just as the second law predicts.

Self-organization is closely related to nonlinear mechanics, chaos, and the geometry of fractals. It has important applications in fields ranging from ecology to computer science to aeronautical engineering. For example, the airflow across a wing gives rise to large-scale turbulence—eddies and whirlpools—in the wake behind an airplane. Their formation affects the aerodynamics of the plane and can also create hazards for following aircraft. Whirlpools are ordered, large-scale macroscopic structures with low entropy, but they are produced from disordered, random collisions of the air molecules.

Self-organizing systems are a very active field of research in both science and engineering. The 1977 Nobel Prize in chemistry was awarded to the Belgian scientist Ilya Prigogine for his studies of *nonequilibrium thermodynamics,* the basic science underlying self-organizing systems. Prigogine and others have shown how energy flow through a system can, when the conditions are right, "bring order out of chaos."

Mathematics Review

Algebra

Using exponents:

$$a^{-x} = \frac{1}{a^x} \qquad a^x a^y = a^{(x+y)} \qquad \frac{a^x}{a^y} = a^{(x-y)} \qquad (a^x)^y = a^{xy}$$

$$a^0 = 1 \qquad a^1 = a \qquad a^{1/n} = \sqrt[n]{a}$$

Fractions:

$$\left(\frac{a}{b}\right)\left(\frac{c}{d}\right) = \frac{ac}{bd} \qquad \frac{a/b}{c/d} = \frac{ad}{bc} \qquad \frac{1}{1/a} = a$$

Logarithms:

If $a = e^x$, then $\ln(a) = x$ $\qquad \ln(e^x) = x \qquad\qquad e^{\ln(x)} = x$

$$\ln(ab) = \ln(a) + \ln(b) \qquad\qquad \ln\left(\frac{a}{b}\right) = \ln(a) - \ln(b) \qquad \ln(a^n) = n\ln(a)$$

The expression $\ln(a + b)$ cannot be simplified.

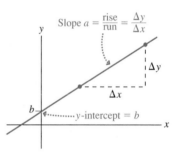

Linear equations: The graph of the equation $y = ax + b$ is a straight line. a is the slope of the graph. b is the y-intercept.

Proportionality: To say that y is proportional to x, written $y \propto x$, means that $y = ax$, where a is a constant. Proportionality is a special case of linearity. A graph of a proportional relationship is a straight line that passes through the origin. If $y \propto x$, then

$$\frac{y_1}{y_2} = \frac{x_1}{x_2}$$

Quadratic equation: The quadratic equation $ax^2 + bx + c = 0$ has the two solutions $x = \dfrac{-b \pm \sqrt{b^2 - 4ac}}{2a}$.

Geometry and Trigonometry

Area and volume:

Rectangle

$A = ab$

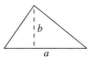

Rectangular box

$V = abc$

Triangle

$A = \frac{1}{2}ab$

Right circular cylinder

$V = \pi r^2 l$

Circle

$C = 2\pi r$

$A = \pi r^2$

Sphere

$A = 4\pi r^2$

$V = \frac{4}{3}\pi r^3$

Arc length and angle:
The angle θ in radians is defined as $\theta = s/r$.

The arc length that spans angle θ is $s = r\theta$.

2π rad $= 360°$

Right triangle:
Pythagorean theorem $\quad c = \sqrt{a^2 + b^2}$ or $a^2 + b^2 = c^2$

$$\sin\theta = \frac{b}{c} = \frac{\text{far side}}{\text{hypotenuse}} \qquad \theta = \sin^{-1}\left(\frac{b}{c}\right)$$

$$\cos\theta = \frac{a}{c} = \frac{\text{adjacent side}}{\text{hypotenuse}} \qquad \theta = \cos^{-1}\left(\frac{a}{c}\right)$$

$$\tan\theta = \frac{b}{a} = \frac{\text{far side}}{\text{adjacent side}} \qquad \theta = \tan^{-1}\left(\frac{b}{a}\right)$$

General triangle:
$\alpha + \beta + \gamma = 180° = \pi$ rad

Law of cosines $c^2 = a^2 + b^2 - 2ab\cos\gamma$

Identities:

$$\tan\alpha = \frac{\sin\alpha}{\cos\alpha} \qquad\qquad \sin^2\alpha + \cos^2\alpha = 1$$

$$\sin(-\alpha) = -\sin\alpha \qquad\qquad \cos(-\alpha) = \cos\alpha$$

$$\sin(\alpha \pm \beta) = \sin\alpha\cos\beta \pm \cos\alpha\sin\beta \qquad \cos(\alpha \pm \beta) = \cos\alpha\cos\beta \mp \sin\alpha\sin\beta$$

$$\sin(2\alpha) = 2\sin\alpha\cos\alpha \qquad\qquad \cos(2\alpha) = \cos^2\alpha - \sin^2\alpha$$

$$\sin(\alpha \pm \pi/2) = \pm\cos\alpha \qquad\qquad \cos(\alpha \pm \pi/2) = \mp\sin\alpha$$

$$\sin(\alpha \pm \pi) = -\sin\alpha \qquad\qquad \cos(\alpha \pm \pi) = -\cos\alpha$$

Expansions and Approximations

Binomial expansion:
$$(1 + x)^n = 1 + nx + \frac{n(n-1)}{2}x^2 + \cdots$$

Binomial approximation:
$$(1 + x)^n \approx 1 + nx \quad \text{if} \quad x \ll 1$$

Trigonometric expansions:
$$\sin\alpha = \alpha - \frac{\alpha^3}{3!} + \frac{\alpha^5}{5!} - \frac{\alpha^7}{7!} + \cdots \quad \text{for } \alpha \text{ in rad}$$

$$\cos\alpha = 1 - \frac{\alpha^2}{2!} + \frac{\alpha^4}{4!} - \frac{\alpha^6}{6!} + \cdots \quad \text{for } \alpha \text{ in rad}$$

Small-angle approximation:
If $\alpha \ll 1$ rad, then $\sin\alpha \approx \tan\alpha \approx \alpha$ and $\cos\alpha \approx 1$.

The small-angle approximation is excellent for $\alpha < 5°$ (≈ 0.1 rad) and generally acceptable up to $\alpha \approx 10°$.

Calculus

The letters a and n represent constants in the following derivatives and integrals.

Derivatives

$$\frac{d}{dx}(a) = 0$$

$$\frac{d}{dx}(ax) = a$$

$$\frac{d}{dx}\left(\frac{a}{x}\right) = -\frac{a}{x^2}$$

$$\frac{d}{dx}(ax^n) = anx^{n-1}$$

$$\frac{d}{dx}\left(\ln(ax)\right) = \frac{1}{x}$$

$$\frac{d}{dx}(e^{ax}) = ae^{ax}$$

$$\frac{d}{dx}\left(\sin(ax)\right) = a\cos(ax)$$

$$\frac{d}{dx}\left(\cos(ax)\right) = -a\sin(ax)$$

Integrals

$$\int x\,dx = \frac{1}{2}x^2$$

$$\int x^2\,dx = \frac{1}{3}x^3$$

$$\int \frac{1}{x^2}\,dx = -\frac{1}{x}$$

$$\int x^n\,dx = \frac{x^{n+1}}{n+1} \quad n \neq -1$$

$$\int \frac{dx}{x} = \ln x$$

$$\int \frac{dx}{a+x} = \ln(a+x)$$

$$\int \frac{x\,dx}{a+x} = x - a\ln(a+x)$$

$$\int \frac{dx}{\sqrt{x^2 \pm a^2}} = \ln\left(x + \sqrt{x^2 \pm a^2}\right)$$

$$\int \frac{x\,dx}{\sqrt{x^2 \pm a^2}} = \sqrt{x^2 \pm a^2}$$

$$\int \frac{dx}{x^2 + a^2} = \frac{1}{a}\tan^{-1}\left(\frac{x}{a}\right)$$

$$\int \frac{dx}{(x^2 + a^2)^2} = \frac{1}{2a^3}\tan^{-1}\left(\frac{x}{a}\right) + \frac{x}{2a^2(x^2 + a^2)}$$

$$\int \frac{dx}{(x^2 \pm a^2)^{3/2}} = \frac{\pm x}{a^2\sqrt{x^2 \pm a^2}}$$

$$\int \frac{x\,dx}{(x^2 \pm a^2)^{3/2}} = -\frac{1}{\sqrt{x^2 \pm a^2}}$$

$$\int e^{ax}\,dx = \frac{1}{a}e^{ax}$$

$$\int xe^{-x}\,dx = -(x+1)e^{-x}$$

$$\int x^2 e^{-x}\,dx = -(x^2 + 2x + 2)e^{-x}$$

$$\int \sin(ax)\,dx = -\frac{1}{a}\cos(ax)$$

$$\int \cos(ax)\,dx = \frac{1}{a}\sin(ax)$$

$$\int \sin^2(ax)\,dx = \frac{x}{2} - \frac{\sin(2ax)}{4a}$$

$$\int \cos^2(ax)\,dx = \frac{x}{2} + \frac{\sin(2ax)}{4a}$$

$$\int_0^\infty x^n e^{-ax}\,dx = \frac{n!}{a^{n+1}}$$

$$\int_0^\infty e^{-ax^2}\,dx = \frac{1}{2}\sqrt{\frac{\pi}{a}}$$

Periodic Table of Elements

Period																		

Key:

Atomic number — 27
Co — Symbol
Atomic mass — 58.9

Transition elements

Inner transition elements

An atomic mass in brackets is that of the longest-lived isotope of an element with no stable isotopes.

Period 1

1 H 1.0 — 2 He 4.0

Period 2

3 Li 6.9 — 4 Be 9.0 — 5 B 10.8 — 6 C 12.0 — 7 N 14.0 — 8 O 16.0 — 9 F 19.0 — 10 Ne 20.2

Period 3

11 Na 23.0 — 12 Mg 24.3 — 13 Al 27.0 — 14 Si 28.1 — 15 P 31.0 — 16 S 32.1 — 17 Cl 35.5 — 18 Ar 39.9

Period 4

19 K 39.1 — 20 Ca 40.1 — 21 Sc 45.0 — 22 Ti 47.9 — 23 V 50.9 — 24 Cr 52.0 — 25 Mn 54.9 — 26 Fe 55.8 — 27 Co 58.9 — 28 Ni 58.7 — 29 Cu 63.5 — 30 Zn 65.4 — 31 Ga 69.7 — 32 Ge 72.6 — 33 As 74.9 — 34 Se 79.0 — 35 Br 79.9 — 36 Kr 83.8

Period 5

37 Rb 85.5 — 38 Sr 87.6 — 39 Y 88.9 — 40 Zr 91.2 — 41 Nb 92.9 — 42 Mo 95.9 — 43 Tc [98] — 44 Ru 101.1 — 45 Rh 102.9 — 46 Pd 106.4 — 47 Ag 107.9 — 48 Cd 112.4 — 49 In 114.8 — 50 Sn 118.7 — 51 Sb 121.8 — 52 Te 127.6 — 53 I 126.9 — 54 Xe 131.3

Period 6

55 Cs 132.9 — 56 Ba 137.3 — 71 Lu 175.0 — 72 Hf 178.5 — 73 Ta 180.9 — 74 W 183.9 — 75 Re 186.2 — 76 Os 190.2 — 77 Ir 192.2 — 78 Pt 195.1 — 79 Au 197.0 — 80 Hg 200.6 — 81 Tl 204.4 — 82 Pb 207.2 — 83 Bi 209.0 — 84 Po [209] — 85 At [210] — 86 Rn [222]

Period 7

87 Fr [223] — 88 Ra [226] — 103 Lr [262] — 104 Rf [265] — 105 Db [268] — 106 Sg [271] — 107 Bh [272] — 108 Hs [270] — 109 Mt [276] — 110 Ds [281] — 111 Rg [280] — 112 Cn [285] — 113 — 114 — 115 — 116 — 117 — 118

Lanthanides 6

57 La 138.9 — 58 Ce 140.1 — 59 Pr 140.9 — 60 Nd 144.2 — 61 Pm 144.9 — 62 Sm 150.4 — 63 Eu 152.0 — 64 Gd 157.3 — 65 Tb 158.9 — 66 Dy 162.5 — 67 Ho 164.9 — 68 Er 167.3 — 69 Tm 168.9 — 70 Yb 173.0

Actinides 7

89 Ac [227] — 90 Th 232.0 — 91 Pa 231.0 — 92 U 238.0 — 93 Np [237] — 94 Pu [244] — 95 Am [243] — 96 Cm [247] — 97 Bk [247] — 98 Cf [251] — 99 Es [252] — 100 Fm [257] — 101 Md [258] — 102 No [259]

ActivPhysics OnLine™ Activities (MP) www.masteringphysics.com

The following list gives the activity numbers and titles of the ActivPhysics activities available in the Pearson eText and the Study Area of MasteringPhysics, followed by the corresponding textbook page references.

1.1	Analyzing Motion Using Diagrams 16	
1.2	Analyzing Motion Using Graphs 46	
1.3	Predicting Motion from Graphs 46	
1.4	Predicting Motion from Equations 49	
1.5	Problem-Solving Strategies for Kinematics 49	
1.6	Skier Races Downhill 49	
1.7	Balloonist Drops Lemonade 52	
1.8	Seat Belts Save Lives 49	
1.9	Screeching to a Halt 49	
1.10	Pole-Vaulter Lands 52	
1.11	Car Starts, Then Stops 49	
1.12	Solving Two-Vehicle Problems 49	
1.13	Car Catches Truck 49	
1.14	Avoiding a Rear-End Collision 49	
2.1.1	Force Magnitudes 141	
2.1.2	Skydiver 141	
2.1.3	Tension Change 141	
2.1.4	Sliding on an Incline 141	
2.1.5	Car Race 141	
2.2	Lifting a Crate 141	
2.3	Lowering a Crate 141	
2.4	Rocket Blasts Off 141	
2.5	Truck Pulls Crate 149	
2.6	Pushing a Crate Up a Wall 149	
2.7	Skier Goes Down a Slope 155	
2.8	Skier and Rope Tow 155	
2.9	Pole-Vaulter Vaults 155	
2.10	Truck Pulls Two Crates 181	
2.11	Modified Atwood Machine 181	
3.1	Solving Projectile Motion Problems 93	
3.2	Two Balls Falling 93	
3.3	Changing the x-Velocity 93	
3.4	Projectile x- and y-Accelerations 93	
3.5	Initial Velocity Components 93	
3.6	Target Practice I 93	
3.7	Target Practice II 93	
4.1	Magnitude of Centripetal Acceleration 102, 194	
4.2	Circular Motion Problem Solving 207	
4.3	Cart Goes Over Circular Path 207	
4.4	Ball Swings on a String 207	
4.5	Car Circles a Track 197	
4.6	Satellites Orbit 366	
5.1	Work Calculations 282	
5.2	Upward-Moving Elevator Stops 297	
5.3	Stopping a Downward-Moving Elevator 297	
5.4	Inverse Bungee Jumper 297	
5.5	Spring-Launched Bowler 297	
5.6	Skier Speed 297	
5.7	Modified Atwood Machine 297	
6.1	Momentum and Energy Change 248	
6.2	Collisions and Elasticity 265	
6.3	Momentum Conservation and Collisions 230	
6.4	Collision Problems 230	
6.5	Car Collision: Two Dimensions 297	
6.6	Saving an Astronaut 230	
6.7	Explosion Problems 230	
6.8	Skier and Cart 297	
6.9	Pendulum Bashes Box 297	
6.10	Pendulum Person-Projectile Bowling 230	
7.1	Calculating Torques 322	
7.2	A Tilted Beam: Torques and Equilibrium 330	
7.3	Arm Levers 330	
7.4	Two Painters on a Beam 330	
7.5	Lecturing from a Beam 330	

7.6	Rotational Inertia 319	
7.7	Rotational Kinematics 105	
7.8	Rotoride: Dynamics Approach 327	
7.9	Falling Ladder 327	
7.10	Woman and Flywheel Elevator: Dynamics Approach 327	
7.11	Race Between a Block and a Disk 335	
7.12	Woman and Flywheel Elevator: Energy Approach 318	
7.13	Rotoride: Energy Approach 318	
7.14	Ball Hits Bat 342	
8.1	Characteristics of a Gas 509	
8.2	Maxwell-Boltzmann Distribution: Conceptual Analysis 509	
8.3	Maxwell-Boltzmann Distribution: Quantitative Analysis 509	
8.4	State Variables and Ideal Gas Law 460	
8.5	Work Done by a Gas 475	
8.6	Heat, Internal Energy, and First Law of Thermodynamics 489	
8.7	Heat Capacity 487	
8.8	Isochoric Process 478	
8.9	Isobaric Process 478	
8.10	Isothermal Process 478	
8.11	Adiabatic Process 489	
8.12	Cyclic Process: Strategies 535	
8.13	Cyclic Process: Problems 535	
8.14	Carnot Cycle 542	
9.1	Position Graphs and Equations 386	
9.2	Describing Vibrational Motion 386	
9.3	Vibrational Energy 384	
9.4	Two Ways to Weigh Young Tarzan 390	
9.5	Ape Drops Tarzan 390	
9.6	Releasing a Vibrating Skier I 393	
9.7	Releasing a Vibrating Skier II 393	
9.8	One- and Two-Spring Vibrating Systems 393	
9.9	Vibro-Ride 393	
9.10	Pendulum Frequency 392	
9.11	Risky Pendulum Walk 392	
9.12	Physical Pendulum 392	
10.1	Properties of Mechanical Waves 566	
10.2	Speed of Waves on a String 563	
10.3	Speed of Sound in a Gas 574	
10.4	Standing Waves on Strings 595	
10.5	Tuning a Stringed Instrument: Standing Waves 602	
10.6	String Mass and Standing Waves 595	
10.7	Beats and Beat Frequency 615	
10.8	Doppler Effect: Conceptual Introduction 580	
10.9	Doppler Effect: Problems 580	
10.10	Complex Waves: Fourier Analysis	
11.1	Electric Force: Coulomb's Law 733	
11.2	Electric Force: Superposition Principle 733	
11.3	Electric Force Superposition Principle (Quantitative) 733	
11.4	Electric Field: Point Charge 740	
11.5	Electric Field Due to a Dipole 755	
11.6	Electric Field: Problems 755	
11.7	Electric Flux 785	
11.8	Gauss's Law 791	
11.9	Motion of a Charge in an Electric Field: Introduction 767	
11.10	Motion in an Electric Field: Problems 767	
11.11	Electric Potential: Qualitative Introduction 819	

11.12	Electric Potential, Field, and Force 844	
11.13	Electrical Potential Energy and Potential 844	
12.1	DC Series Circuits (Qualitative) 898	
12.2	DC Parallel Circuits 904	
12.3	DC Circuit Puzzles 906	
12.4	Using Ammeters and Voltmeters 906	
12.5	Using Kirchhoff's Laws 906	
12.6	Capacitance 910	
12.7	Series and Parallel Capacitors 910	
12.8	RC Circuit Time Constants 910	
13.1	Magnetic Field of a Wire 928	
13.2	Magnetic Field of a Loop 930	
13.3	Magnetic Field of a Solenoid 938	
13.4	Magnetic Force on a Particle 940	
13.5	Magnetic Force on a Wire 947	
13.6	Magnetic Torque on a Loop 948	
13.7	Mass Spectrometer 942	
13.8	Velocity Selector 942	
13.9	Electromagnetic Induction 975	
13.10	Motional emf 975	
14.1	The RL Circuit 991	
14.2	The RLC Oscillator 1043	
14.3	The Driven Oscillator 1043	
15.1	Reflection and Refraction 661	
15.2	Total Internal Reflection 661	
15.3	Refraction Applications 661	
15.4	Plane Mirrors 659	
15.5	Spherical Mirrors: Ray Diagrams 682	
15.6	Spherical Mirror: The Mirror Equation 682	
15.7	Spherical Mirror: Linear Magnification 682	
15.8	Spherical Mirror: Problems 682	
15.9	Thin-Lens Ray Diagrams 673	
15.10	Converging Lens Problems 679	
15.11	Diverging Lens Problems 679	
15.12	Two-Lens Optical Systems 679	
16.1	Two-Source Interference: Introduction 631	
16.2	Two-Source Interference: Qualitative Questions 631	
16.3	Two-Source Interference: Problems 631	
16.4	The Grating: Introduction and Qualitative Questions 636	
16.5	The Grating: Problems 636	
16.6	Single-Slit Diffraction 637	
16.7	Circular Hole Diffraction 640	
16.8	Resolving Power 709	
16.9	Polarization 1024	
17.1	Relativity of Time 1074	
17.2	Relativity of Length 1078	
17.3	Photoelectric Effect 1128	
17.4	Compton Scattering	
17.5	Electron Interference 1135	
17.6	Uncertainty Principle 1169	
17.7	Wave Packets 1169	
18.1	The Bohr Model 1138	
18.2	Spectroscopy 1104, 1232	
18.3	The Laser 1238	
19.1	Particle Scattering 1114	
19.2	Nuclear Binding Energy 1253	
19.3	Fusion 836	
19.4	Radioactivity 1263	
19.5	Particle Physics 1266	
20.1	Potential Energy Diagrams 1182	
20.2	Particle in a Box 1185	
20.3	Potential Wells 1194	
20.4	Potential Barriers 1206	

PhET Simulations (MP) www.masteringphysics.com

The following list gives the titles of the PhET simulations available in the Pearson eText and in the Study Area of MasteringPhysics, with the corresponding textbook section and page references.

1.3 Vector Addition 7, 71, 75
1.8 Estimation 27
2.1 Equation Grapher 36
2.2 Calculus Grapher 40, 44
2.3 *The Moving Man 42
4.1 Motion in 2D 86, 102
4.1 Maze Game 86, 88
4.1 Ladybug Motion in 2D 86
4.3 *Projectile Motion 93
4.5 Ladybug Revolution 99
5.5 *Forces in 1 Dimension 126, 127
5.7 *The Ramp 130
6.1 Molecular Motors 139
6.2 Lunar Lander, *The Ramp 142
6.4 *Forces in 1 Dimension, Friction 149
6.6 *The Ramp 155
8.2 Ladybug Revolution, Motion in 2D 194
8.3 My Solar System 199
8.5 Ladybug Motion in 2D 205
9.5 Lunar Lander 236
10.2 *Energy Skate Park 248, 261
10.5 *Masses & Springs 257
10.6 Stretching DNA 261
11.2 *The Ramp 280
11.7 States of Matter 292
12.1 Ladybug Revolution 313
12.3 Torque 317, 322, 340
13.3 Gravity Force Lab 357
13.6 My Solar System 366
14.1 Motion in 2D 379
14.5 *Masses & Springs 390
14.6 *Pendulum Lab 392, 395
15.1 States of Matter 408
15.2 Gas Properties 411
15.4 Balloons & Buoyancy 421
15.5 *Energy Skate Park 427
16.1 States of Matter 445, 451
16.5 Gas Properties 453, 457
17.2 States of Matter 473, 476
17.8 Blackbody Spectrum 493
17.8 The Greenhouse Effect 494
18.1 Gas Properties 503
18.6 Reversible Reactions 518
20.1 Waves on a String, *Wave Interference 561
20.2 *Wave Interference 565, 574
20.3 Waves on a String 570, 574
20.5 Sound, Radio Waves and EM Fields 574
21.1 Fourier: Making Waves, Waves on a String 592
21.3 Waves on a String 595
21.4 *Wave Interference 599
22.2 *Wave Interference 630, 637

23.5 Color Vision 667
23.6 *Geometric Optics 672, 674
24.4 *Geometric Optics 703
25.3 Conductivity 727
25.3 Balloons and Static Electricity, John Travoltage 728, 731
25.5 *Charges and Fields 740
26.2 *Charges and Fields 752, 754, 759, 760, 764
26.6 Electric Field of Dreams, Electric Field Hockey 767
26.6 Microwaves, Optical Tweezers and Applications 767
28.4 *Charges and Fields 819, 828
29.2 Battery Voltage 843
30.2 Signal Circuit 871
30.5 Resistance in a Wire, Ohm's Law 880
30.5 Battery-Resistor Circuit 883
31.2 *Circuit Construction (DC Only) 893, 898, 904, 906, 910
31.3 Battery-Resistor Circuit 896
32.1 Magnet and Compass 922
32.5 Magnets and Electromagnets 931, 939
33.1 *Faraday's Electromagnetic Lab, Faraday's Law 963
33.3 *Faraday's Electromagnetic Lab 969
33.4 Faraday's Law 972
33.7 Generator, *Faraday's Electromagnetic Lab 983
33.8 *Circuit Construction (AC + DC) 986
34.5 Radio Waves & Electromagnetic Fields 1016, 1023
35.1 *Circuit Construction (AC + DC) 1034, 1036, 1041, 1043
36.10 Nuclear Fission 1094
37.1 Blackbody Spectrum 1104
37.4 Neon Lights and Other Discharge Lamps 1108
37.6 Rutherford Scattering 1114
38.1 Photoelectric Effect 1128
38.6 Models of the Hydrogen Atom 1141
38.7 Neon Lights and Other Discharge Lamps 1146
39.1 Wave Interference 1157
39.1 Plinko Probability 1158
39.1 Quantum Wave Interference 1159
40.6 Quantum Bound States 1194
40.9 Double Wells and Covalent Bonds 1206
40.10 Quantum Tunneling and Wave Packets 1206
41.1 Models of the Hydrogen Atom 1217
41.3 Stern-Gerlach Experiment 1223
41.6 Neon Lights and Other Discharge Lamps 1234
41.8 Lasers 1238
42.5 Radioactive Dating Game 1260
42.6 Alpha Decay, Beta Decay 1264
42.7 Simplified MRI 1270

*Indicates an associated tutorial is available in the MasteringPhysics Item Library.

Answers

Answers to Odd-Numbered Exercises and Problems

Chapter 16

1. $1900\ cm^3$
3. $8.0\ cm$
5. 4.8×10^{23} atoms
7. a. 6.02×10^{28} atoms/m^3 b. 3.28×10^{28} atoms/m^3
9. $6.8\ cm^3$
11. $-127°F = 88°C = 185\ K$; $136°F = 58°C = 331\ K$
13. a. $171°Z$ b. $671°C = 944\ K$
15. a. 2 b. Unchanged
17. a. $1.27V_0$ b. $2V_0$
19. 2.4×10^{22} molecules
21. $7.4\ kg/m^3$
23. a. $V_2 = V_1$ b. $T_2 = \dfrac{T_1}{3}$
25. $2.6\ atm$
27. a. $9500\ kPa$
   b.

29. a. $48\ atm$
31. a. Isobaric   b. $118°C$ c. 9.35×10^{-3} mol
33. $0.228\ nm$
35. 3.3×10^{26} protons
37. 1.1×10^{15} particles/m^3
39. $380\ K = 107°C$
41. $1.8\ g$
43. $\dfrac{3}{2}T_0$
45. $2.4\ m$
47. $35\ psi$
49. $155\ cm^3$
51. $24\ cm$
53. No
55.

57. a. $880\ kPa$ b. $T_2 = 323°C$, $T_3 = -49°C$, $T_4 = 398°C$
59. a. $T_1 = 122\ K$, $T_2 = 366\ K$ b. Isobaric   c. $3\ atm$
61. $2364°C$
63. a. $4.0\ atm$, $-73°C$
   b.

65.   b.

   c. $6\ atm$
67.   b.   c. $417\ cm^3$

69. a. $23\ cm$ b. $7.5\ cm$
71. $93\ cm^3$
73. a. $4.0 \times 10^5\ Pa$ b. Irreversible

Chapter 17

1. 60 J
3. 200 cm³
5.
7.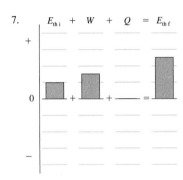
9. 60 J
11. 52 kJ
13. a. 36°C b. 3000 J
15. 0.98 g
17. 272°C, 522°F
19. Iron
21. a. 31 J b. 60°C
23. 2.5 kJ
25. a. 1.9×10^{-3} m³ b. 74°C
27. 16 kW
29. 230 W
31. 16 kJ
33. 15 m
35. 6.6 h
37. 12 J/s
39. −56°C
41. Aluminum
43. 650 J/kg K
45. a. 2.0 kJ/kg K b. 2.7 kJ/kg K c. −20°C, 40°C
    d. 4.0×10^4 J/kg, 1.2×10^5 J/kg
47. 2.4×10^6 L
49. a. 5.5 kJ b. 3.4 kJ
    c. p (atm)
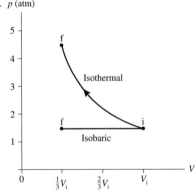

51. a. 350 Pa b. 4.9×10^{20} c. 110°C d. 26 cm e. −0.57 J
53. a. 3.1 atm b. 9.7 L
    c. p (atm)

55. a. T_1 b. $-nRT_1\ln2$ c. $nRT_1\ln2$
59. −330 J, 0 J
61. $\gamma = 1.29$
63. a. 0.15 kJ b. −91 J
65. a.

Point	p (atm)	T (°C)	V (cm³)
1	1.0	133	1000
2	5.0	1757	1000
3	1.0	1757	5000

   b. $W_{1\to2} = 0$, $W_{2\to3} = -815$ J, $W_{3\to1} = 405$ J c. $Q_{1\to2} = 609$ J,
   $Q_{2\to3} = 815$ J, $Q_{3\to1} = -1.01$ kJ
67. 28°C
69. a. 5.5 kK b. 0 J c. 54 kJ d. 20
    e. p (atm)
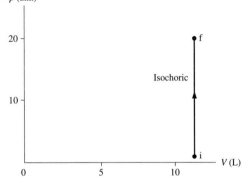

71. 110°C
73. −18°C
75. b. 217°C
77. a.

Point	p (atm)	T (°C)	V (cm³)
1	3.0	946	1000
2	1.0	946	3000
3	0.48	310	3000

   b. −334 J, 0 J, 239 J c. 334 J, −239 J, 0 J
79. 15 atm
81. 150 J

Chapter 18

1. 5.5×10^{24}
3. 0.023 Pa
5. a. 300 nm b. 600 nm
7. 13 cm
9. a. $(0\hat{i} + 0\hat{j})$ b. 57 m/s c. 60 m/s
11. a. 289 K b. 200 kPa

13. 6.5×10^{25} s^{-1}

15. 283 m/s

17. $-246°$C

19. 300 m/s

21. 0.43 cm/s

23. a. Doubles   b. $\sqrt{2}$ c. Same

25. a. 4.1×10^{-16} J   b. 7.0×10^5 m/s

27. 580 m/s

29. 3.6×10^7 J

31. 93 kJ

33. a. 3.80×10^5 J   b. 2.25×10^{-9} m   c. 0 J

35. 5000 J

37. 61

39. a. Helium   b. 1370 m/s   c. 1.86 μm

41. a. 4×10^{-22} atm   b. 270 m/s   c. 2.5×10^5 m

43. 1.004

45. 1.9×10^4 Pa

47. 29 J/mol K

49. a. $(E_{He})_i = 1900$ J, $(E_O)_i = 3100$ J
    b. $(E_{He})_f = 2700$ J, $(E_O)_f = 2300$ J
    c. 850 J from oxygen to helium   d. 436 K

51. 7

55. a. Increase factor of 2   b. Increase by factor of 4
    c. Increase by factor of 4   d. Same

57. a. 4   b. 1   c. 16

59. a. 141,000 K   b. 10,100 K

61. a. 2.0×10^6 J   b. 4.8×10^{-6} c. 0.0013 K

63. $\dfrac{15n + 3}{2} p_i V_i$

65. c. 436 K; 850 J is transferred from oxygen to helium

Chapter 19

1. a. 250 J   b. 150J

3. a. 0.27   b. 15 kJ

5. a. 200 J   b. 250 J

7. 96,000

9.

	ΔE_{th}	W_s	Q
A	+	0	+
B	−	+	0
C	0	+	+
D	−	−	−

11. 40 J

13. a. 30 J, 0.15 kJ   b. 0.21

15. 285 J

17. 0.24

19. a. (b)   b. (a)

21. 7°C

23. a. 25%   b. 232°C

25. 135°C

27. 1.7

29. a. 60 J   b. -23°C

31. 1.7 MJ

37. 8.3%

39. 47°C

41. 218

43. 8.57 J

45. No

47. a. 48 m   b. 32%

49. 37%

51. a. 5.0 kW   b. 1.7

53. a.

	W_s (J)	Q (J)	ΔE_{th}
$1 \to 2$	3.04	16.97	13.93
$2 \to 3$	0	−10.13	−10.13
$3 \to 1$	−1.52	−5.32	−3.80
Net	1.52	1.52	0

b. 9.0%   c. 13 W

55. a.

	W_s (kJ)	Q (kJ)	ΔE_{th} (kJ)
$1 \to 2$	0.991	2.476	1.486
$2 \to 3$	0	−1.693	−1.693
$3 \to 1$	−0.207	0	0.207
Net	0.783	0.783	0

b. 0.32

57. a. $p_1 = 100$ kPa $V_1 = 2690$ cm^3 $T_1 = 269$ K

b.

	ΔE_{th} (J)	W_s (J)	Q (J)
$1 \to 2$	327	−327	0
$2 \to 3$	0	555	555
$3 \to 1$	−458	0	−458
Net	−131	228	97

c. 17%

59. a.

	p (atm)	T (K)	V (cm^3)
1	1.0	406	1000
2	5.0	2030	1000
3	1.0	2030	5000

b. 29%   c. 80%

61. a. $T_1 = 1620$ K $T_2 = 2407$ K $T_3 = 6479$ K

b.

	ΔE_{th} (J)	W_s (J)	Q (J)
$1 \to 2$	327	−327	0
$2 \to 3$	1692	677	2369
$3 \to 1$	−2019	0	−2019
Net	0	350	350

c. 15%

63. a. $W_{net} = 350$ J   b. $\eta = 0.24$

65. b. 1.1×10^3 °C

67. b. $Q_C = 80$ J

69. b. 10 J   c. 0.13

Credits

PART IV OPENER
Page **442:** ImageState/Alamy.

CHAPTER 16
Page **444** Top: Frans Lanting Studio/Alamy. Page **444** Middle: NASA. Page **444** Bottom: Fotolia. Page **447:** Richard Megna/Fundamental Photographs. Page **449** Both: David Young-Wolff/PhotoEdit, Inc. Page **452:** Vibe Images/Alamy.

CHAPTER 17
Page **469** Top: Ted Kinsman/Peter Arnold Inc./Photolibrary. Page **469** Bottom left: Shutterstock. Page **469** Bottom middle: iStockphoto. Page **469** Bottom right: Clouds Hill Imaging Ltd./Photo Researchers, Inc. Page **471:** Alexey Dudoladov/Getty Images RF. Page **477:** Gordon Saunders/Shutterstock. Page **482:** Getty Images/Stocktrek Images. Page **491** Left: Cn Boon/Alamy. Page **491** Middle left: Andrew Davidhazy. Page **491** Middle right: Pascal Goetgheluck/Photo Researchers, Inc. Page **491** Right: Corbis RF. Page **492:** Science Photos/Alamy. Page **493:** Johns Hopkins University Applied Physics Laboratory.

CHAPTER 18
Page **502** Top: Jose Mendes/Fotolia. Page **502** Bottom: Rick Lord/Shutterstock. Page **519:** Getty Images – Photodisc.

CHAPTER 19
Page **526** Top: imagebroker.net/SuperStock. Page **526** Bottom left: Peter Bowater/Alamy. Page **526** Bottom right: Shutterstock. Page **529:** Peter Bowater/Alamy. Page **532:** Shutterstock. Page **536:** Malcolm Fife/Getty Images - Photodisc.

PART IV SUMMARY
Page **557:** AGE Fotostock/SuperStock.

Index

A

Absolute temperature, 450
Absolute zero, 449, 450, 452
Absorbed dose, 1268
Absorption
    excitation by, 1232–33
    in hydrogen, 1233
    of light, 1234, 1235
    in sodium, 1233
Absorption spectra, 1105, 1184, 1190, 1203
Acceleration, 1, 86–87
    angular, 103–05, 205, 313–14
    average, 13, 86, 88, 89
    centripetal, 102–03, 194, 200, 205
    constant, 90–91, 124
    force and, 124–25, 133, 141
    free-fall, 52
    Galilean transformation of, 1005
    instantaneous, 58–60, 89–90
    nonuniform, 58–59
    sign of, 16–17, 61
    tangential, 105–06, 205
    in two dimensions, 192–93
    in uniform circular motion, 102–03
Acceleration constraints, 174–75, 184
Acceleration vectors, 13–14, 86–87
AC circuits, 1031–55
    AC sources, 1034
    capacitor circuits, 1036–38
    inductor circuits, 1041–42
    phasors, 1024
    power in, 1046–49
    RC filter circuits, 1038–41
    resistor circuits, 1034–36
    series *RLC* circuits, 1042–46
Accommodation, 701, 702
Action/reaction pair, 168. *See also* Newton's
    third law of motion
    identifying, 169–72
    massless spring approximation and, 179
    propulsion, 171
Activity of radioactive sample, 1261
Adiabatic processes, 479, 488–91
Adiabats, 489
Agent of force, 117
Air resistance, 51, 121, 128, 153
Airplanes, lift in, 429

Allowed transitions, 1232
Alpha decay, 1264
Alpha particles, 836, 1113–14, 1116, 1258, 1263–64
Alpha rays, 1113, 1249
Alternating current (AC), 983. *See also* AC circuits
Ammeters, 900
Amorphous solids, 445
Ampere (A), 875
Ampère-Maxwell law, 1012–14, 1015, 1019–20
Ampère's law, 934–40, 1011–12
Amplitude, 378, 380, 396, 398–99, 566, 570, 573, 579, 593–94
Amplitude function, 595
Angle of incidence, 658, 661, 662
Angle of reflection, 658
Angle of refraction, 661, 662
Angular acceleration, 103–05, 108, 205, 313–14
Angular displacement, 99
Angular frequency, 379, 381
Angular magnification, 704, 705, 706
Angular momentum, 312, 340–45, 1223–25
    conservation of, 342, 368
    hydrogen atom, 1218–19
    quantization of, 1145
    of a rigid body, 341
Angular momentum vector, 340
Angular position, 98–99, 631
Angular resolution, 709
Angular size, 703
Angular velocity, 99–101, 108, 205, 313–14, 342–45
Angular velocity vector, 337–40
Antennas, 1023
Antibonding molecular orbital, 1206
Antimatter, 1093
Antinodal lines, 612, 630
Antinodes, 594, 599, 600, 601
Antireflection coatings, 608–10
Apertures, 657, 698
Archimedes' principle, 420
Arc length, 99
Area vector, 787, 970
Atmospheric pressure, 411–13, 416–17

Atomic clocks, 23, 1076–77
Atomic magnets, 950
Atomic mass, 292, 447, 1250–51
Atomic mass number (*A*), 447, 1198
Atomic mass unit (u), 447, 1250, 1251, 1254
Atomic model, 120, 264
Atomic number, 1117, 1249
Atomic physics, 1216–47
    electron spin, 1218, 1223–25
    emission spectra, 1231–35
    excited states, 1236–38
    hydrogen atom, 1217–23
    lasers, 1238–42
    multielectron atoms, 1225–28
    periodic table of the elements, 1228–31
Atoms, 217, 453, 1103, 1108. *See also* Electrons; Hydrogen; Nuclear Physics; Nucleus; Protons
    Bohr model of, 1138, 1141–46
    electricity and, 725–26
    hard-sphere model of, 453
    nuclear model of, 1114, 1116–17
    plum-pudding or raisin-cake model of, 1112–13
    shell model of, 1222, 1228–31, 1256–57
    structure of, 725, 729, 1118, 1138, 1249
Avogadro's number, 292, 447–48

B

Ballistic pendulum, 253, 344–45
Balmer formula, 1105
Balmer series, 1105
Bandwidth, 1168
Bar charts, 223–24, 249–50, 295–96
Barometers, 416
Basic energy model, 245, 246–47, 279–80
Batteries, 849–50, 881–82, 892–900. *See also* Circuits
    charge escalator model of, 843–44
    emf of, 843
    ideal, 843
    real, 901–03
    short circuited, 902
    as source of potential, 823, 882
Beats, 615–18
    beat frequency, 617

Before-and-after pictorial representation, 223–24
Bernoulli's equation, 426–30
Beta decay, 1258–59, 1264–66
Beta particles, 1258–59
Binding energy, 1145, 1253–55
Binoculars, 675
Binomial approximation, 762, 1086
Biot-Savart law, 925, 927–28, 1006–07
Blackbody radiation, 493, 1104
Blood pressure, 417–18
Bohr radius, 1143, 1221
Bohr's analysis of hydrogen atom, 1141–46, 1147
Bohr's model of atomic quantization, 1138–41, 1217, 1222
Boiling point, 449, 451, 482–83
Boltzmann's constant, 455, 1197
Bonding molecular orbital, 1206
Boundary conditions
    for standing sound waves, 599–601
    for standing waves on a string, 596–99
    for wave functions, 1183–87
Bound states, 1194, 1206
Bound system, 369, 1253
Brayton cycle, 536–38
Breakdown field strength, 748
Bulk modulus, 432–33
Bulk properties, 444
Buoyancy, 419–23
Buoyant force, 419

C

Calories, 477
Calorimetry, 483–85
Camera obscura, 657
Cameras, 695–99
    controlling exposure, 698–99
    focusing, 697
    zoom lenses, 697
Capacitance, 850, 851, 899
Capacitative resistance, 1037
Capacitor circuits, 1036–38
Capacitors
    and capacitance, 849–50
    charging, 764, 912
    current and voltage in AC circuits, 1036
    dielectrics in, 855–58
    discharging, 868, 870, 912
    energy stored in, 854–55
    parallel and series capacitors, 851–85
    parallel-plate capacitors, 764–66
    RC circuits, 909–12
Carbon dating, 1250, 1264
Carnot cycle, 542–46
Carnot engine, 542–46

Cathode rays, 1106–08
Cathode-ray tube (CRT) devices, 768, 1107, 1108
Causal influence, 1089–90
CCD (charge-coupled device), 696, 699
Celsius scale, 449, 452
Center of mass, 312, 394
    in balance and stability, 333
    rotation around, 314–17
Central maximum, 630, 631, 636–37, 639, 640–41
Centrifugal force, 202–03
Centripetal acceleration, 102–03, 194, 200, 205
Charge carriers, 727, 868–69
Charge density, 757
Charge diagrams, 726
Charge distribution. *See also* Electric fields
    continuous, 756–760, 828–29
    symmetric, 781–83
Charge model, 721–25
Charge polarization, 729–31
Charge quantization, 726
Charges, 719, 722, 725–26, 1056
    atoms and electricity, 725
    on capacitors, 850
    conservation of, 726, 892, 896
    discharging, 724
    fundamental unit of, 725, 1111–12
    like charges, 722
    neutral, 723, 726
    opposite, 722
    positive and negative, 725
    units of, 732–33
Charge-to-mass ratio, 767, 942
Charging, 719, 721
    frictional, 721–23, 726
    by induction, 731
    insulators and conductors, 727–28
    parallel-plate capacitors, 849–50
Chromatic aberration, 707
Circuits, 891–920. *See also* AC circuits; DC circuits
    diagrams, 892
    elements, 892
    energy and power in, 896–98
    grounded circuits, 908–09
    Kirchhoff's laws and, 892–96
    LC, 988–90
    LR, 391–93
    oscillator, 920
    parallel resistors, 903–06
    RC, 909–12
    resistor circuits, 906–08
    series resistors, 898–901
Circular-aperture diffraction, 640–42, 708

Circular motion, 3, 98–107, 108, 216.
    See also Rotational motion
    angular acceleration, 103–07
    dynamics of, 193–99
    fictitious forces, 201–04
    nonuniform, 103–107, 205–08
    orbits, 199–201
    period of, 98, 100–01
    problem-solving strategy, 207
    simple harmonic motion (SHM) and, 381–84
    uniform. *See* Uniform circular motion
Circular waves, 572, 610–15
Classically forbidden region, 1195–96
Classical physics, 1015, 1127–29
    end of, 1118–1129
Clocks
    atomic, 1076
    synchronizing, 1070
    time dilation and, 1075
Closed-cycle devices, 529
Coaxial cable, 961, 1002
Coefficients of friction, 149
    kinetic friction, 149, 206
    rolling friction, 149
    static friction, 148
Coefficients of performance, 532, 544
Cold reservoir, 528, 531
Collisional excitation, 1233–34
Collisions, 219, 221–26, 870, 873–74
    elastic, 232, 265–69, 1065
    inelastic, 232–34, 253–54
    mean free path between, 503–04
    mean time between, 874
    molecular, 503–07, 514–20
    pressure and, 503
Color, 667–70
    chromatic aberration, 707
    human color vision, 700
    in solids, 1234–35
Compasses, 923
Component vectors, 75–77
Compression, 431
    adiabatic, 479, 488–89
    isobaric, 458
    in sound waves, 574, 599
Concave mirrors, 682
Condensation point, 450, 451
Conduction
    electrical, 868–70, 1127
    heat, 491–92
    model of, 873–74
Conductivity
    electrical, 878–80
    thermal, 492
Conductors, 724, 727–31
    charge carriers, 868–69

in electrostatic equilibrium, 799–802, 848–49
Conservation of angular momentum, 342–43
Conservation of charge, 726, 877, 892, 896
Conservation of current, 876–78, 892, 896, 903, 920
Conservation of energy, 219, 294–97, 301, 813–14, 897
  in charge interactions, 820
  in double-slit interference, 633
  in fluid flow, 419, 426
  in relativity, 1093–95
  in simple harmonic motion (SHM), 385–86, 440
Conservation laws, 219–20
Conservation of mass, 219, 235, 1093
Conservation of mechanical energy, 254–55, 270
Conservation of momentum, 219, 220, 226–32, 1065
Conservative forces, 288–89, 294, 811
Constant-pressure (isobaric) process, 458–59
Constant-temperature (isothermal) process, 459–60
Constant-volume gas thermometer, 449
Constant-volume (isochoric) process, 457–59
Constructive interference, 594, 605–07, 609–15, 630–39
Contact forces, 117–22
Continuity, equation of, 424–25
Continuous spectra, 1103–04
Contour maps, 614, 823–25
Convection, 491, 492
Converging lenses, 671–73, 702
Convex mirrors, 682
Coordinate systems, 5, 74–77
  displacement and, 8
  right-handed, 339
  rtz, for circular dynamics, 193–94, 209
  tilted axes, 79–80
Correspondence principle, 1191–93
Coulomb (C), 732
Coulomb electric fields, 979
Coulomb's law, 731–32, 733–36, 739, 753, 814, 1006, 1007
Covalent bonds, 1205–06
Critical angle, 333, 665
Critical point, 452
Crookes tubes, 1107
Crossed-field experiment, 1109–10
Crossed polarizers, 1025
Crossover frequency, 1039
Cross product, 338–39, 927

Current, 719, 727, 867–90, 892–94. *See also* Circuits; Electron current; Induced currents
  batteries and, 849–50
  conservation of, 876–78, 892, 896, 903, 920
  creating, 870–73
  displacement, 1011–14
  eddy currents, 968
  magnetic field of, 925–31
  magnetic forces on wires carrying, 946–50
  and magnetism, 923–25
  root-mean-square (rms) current, 1046–47
Current density, 876, 878
Current loops, 930
  forces and torques on, 948–50
  as magnetic dipoles, 932–34
  magnetic field of, 930–31
Curve of binding energy, 1254
Cyclotron, 943–44
Cyclotron frequency, 942
Cyclotron motion, 941–43

D

Damped oscillations, 395–99
Damping constant, 395
Daughter nucleus, 1261, 1263
DC circuits, 1034
de Broglie wavelength, 1135, 1194, 1197, 1204
Decay equation, 1236–38
Decay, exponential, 396, 992
Decay, nuclear, 1263–67
Decay rate, 1237
Decay series, 1267
Decibels (dB), 579–80
Defibrillators, 854, 858
Degrees of freedom, 511–13
Density, 408–09, 421, 756
  mass density, 408, 445
  nuclear density, 1251–52
  number density, 446
  surface charge, 757
Destructive interference, 605, 609–11, 612, 630–39
Deuterium, 1251
Diatomic gases, 447, 453, 487
Diatomic molecules, 264–65, 512–14
Dielectric constant, 857
Dielectric strength, 858
Dielectrics, 855–58
Diesel cycle, 536
Diffraction, 628, 657, 707–09
  circular-aperture, 640–42

  of electrons, 1108, 1135–36
  single-slit, 636–40
Diffraction grating, 634–36
  resolution of, 653
Diffuse reflection, 659
Diodes, 883
Diopter, 701
Dipole moment, 754
Dipoles. *See* Electric dipoles; Magnetic dipoles
Direct current (DC) circuits, 1034. *See also* Circuits
Discharging, 724, 728–29, 870, 909, 910
Discrete spectra, 1105
Disk of charge, electric field of, 761–63
Disordered systems, 518–19
Dispersion, 668–69, 707
Displaced fluid, 420
Displacement, 6–9, 43, 61
  angular, 99
  work and, 281
Displacement current, 1011–14
Displacement vectors, 70–71
Dissipative forces, 289–90, 293–94
Diverging lenses, 671, 675–76, 702
Doppler effect, 580–83
Dose, absorbed, 1268
Dose equivalent, 1268
Dot product of vectors, 284–85, 298
Double-slit experiment, 629–34, 668
Double-slit interference, 629–34, 637–38, 668, 1157
  intensity of, 633–34
  interference fringes, 629–630, 1159
Drag coefficient, 153
Drag force, 121, 152–53
  terminal speed and, 154
Drift speed, electron, 868, 869, 873
Driven oscillations, 398–99
Driving frequency, 398–99
Dynamic equilibrium, 128, 141
Dynamics, 1, 216
  fluid, 423–30
  of nonuniform circular motion, 205–08
  in one dimension, 138–66
  problems, strategy for solving, 142, 159
  of rotational motion, 325–27
  of simple harmonic motion (SHM), 386–89
  in two dimensions, 192–93
  of uniform circular motion, 193–99

E

Earthquakes, 257
Eddy currents, 968

Effective focal length, 697. *See also* Focal length
Effective gravitational force, 146, 202
Efficiency, thermal, 529–30
Einstein, Albert
   explanation of photoelectric effect, 1129–32
   photons and photon model of light, 1132–34
   principle of relativity, 1066–68, 1278
Elasticity, 255, 430–33
   tensile stress, 430–32
   volume stress, 432–33
Elastic collisions, 232, 265–69, 1065
Elastic potential energy, 257–61
Electric charge. *See* Charges
Electric dipoles, 730
   electric field of a dipole, 754–55
   motion in an electric field, 770–72
   potential energy of, 817–18
Electric field lines, 755–56
Electric field strength, 738, 878
Electric fields, 737–42, 750–79, 818–19, 1004, 1012
   changing, producing induced magnetic fields, 981–82
   of charged wires, 924
   of conductors in electrostatic equilibrium, 799–801, 848–49
   of continuous charge distribution, 756–60
   Coulomb and non-Coulomb, 979
   crossed-field experiment, 1109–10
   of disk of charge, 761–63
   of electric dipole, 754–55, 933
   electric potential and, 823, 840
   energy in, 855, 988
   establishing in a wire, 871–72
   finding electric potential from, 840–41
   finding with Gauss's law, 795–99
   Gauss's law, 791–94
   induced, 856, 978–81, 1010, 1013
   insulators in, 856–57
   models of, 751
   motion of charged particle in, 767–69
   motion of dipole in, 770–71
   of plane of charge, 763–64
   of a point charge, 739–40
   of multiple point charges, 752–55
   of parallel-plate capacitor, 764–66, 821–22, 850, 851
   picturing, 755–56
   of ring of charge, 60–61
   of sphere of charge, 764–66
   symmetry of, 781–83, 791
   transformations of, 1005–10

uniform, 766, 811–14
   units, volts per meter, 822
Electric flux, 785–94, 1010, 1012, 1013, 1017, 1019
   of nonuniform electric field, 787–88
Electric force, 121–22, 264–65, 816
Electric potential, 818–838
   batteries and, 843
   in closed circuits or loops, 893
   of conductor in electrostatic equilibrium, 848–49
   electric potential energy vs., 819
   finding from electric field, 840–41
   finding electric field from, 844–48
   inside parallel-plate capacitor, 821–25
   of many charges, 828–30
   of a point charge, 826–27
   sources of, 842–44
Electric potential energy, 811–14
   of a dipole, 817–18
   electric potential vs., 819
   of point charges, 814–17
Electrical oscillators, 1034
Electricity, 719–49, 1056
   charge, 723–26, 1056
   charge model, 721–25
   field model, 736–42
   insulators and conductors, 727–31
   phenomena and theories, 719
Electrodes, 763–64, 765
   capacitance and, 850, 851
   equipotential surface close to, 848
Electromagnetic fields, 1003–16
   forces and, 1004
   Maxwell's equations, 1014–16
   transformation of, 1005–10
Electromagnetic induction, 962–1002
   Faraday's law, 975–78
   induced currents, 963
   induced fields, 978–81
   Lenz's law, 971–74
   motional emf, 964–68
   problem-solving strategy for, 976
Electromagnetic radiation, quantized, 1129
Electromagnetic spectrum, 576
Electromagnetic waves, 561, 575–77, 1016–20, 1066
   Ampère-Maxwell law, 1019–20
   energy and intensity, 1021–22
   generated by antennas, 1023
   light, 628
   Maxwell's theory of, 981–82
   polarization, 1024–26
   radiation pressure, 1022
   speed of, 1020
Electromagnets, 932

Electron capture, 1265
Electron cloud, 725, 1220
Electron configuration, 1227
Electron current, 868–870
Electron spin, 950–51, 952
Electronic ignition, 987
Electrons
   beams, 769
   charge of, 725–26
   charge carriers in metals, 868–69
   diffraction of, 1135
   discovery of, 1108–11
   sea of, 727
   spin of, 1218, 1223–25
Electron volt, 1115
Electroscope, 728–29, 731
Electrostatic constant, 732, 733
Electrostatic equilibrium, 727, 870–71
   conductors in, 799–802, 848–50
Electrostatic forces, and Coulomb's law, 733–36
Electroweak force, 217
Elements, 1117, 1139, 1228–31
emf
   of batteries, 843
   induced emf, 975–78, 983
   motional emf, 964–68
Emission spectrum, 1103–04, 1105, 1234
   hydrogen atom, 1106, 1146–47
Emissivity, 493
Endoscope, 685–86
Energy, 219, 470. *See also* Conservation of energy; Kinetic energy; Mechanical energy; Potential energy; Thermal energy; Work
   allowed energy for quantum mechanics problems, 1187–88
   angular and linear momentum and, 341
   bar charts, 249–50, 295–96, 301
   basic energy model, 245, 246–47, 279–80, 301
   binding energy, 1145, 1253–55
   in circuits, 896–98
   in damped systems, 396–97
   diagrams, 261–65, 813
   in electric fields, 855
   of electromagnetic waves, 1021
   forms of, 245, 246
   ionization energy, 1145
   in magnetic fields, 988
   of photons, 1133–34
   of photoelectrons, 1127–28
   problem-solving strategy, 297, 301
   quantization of, 1136–38, 1184
   relativistic, 1090–95
   rotational, 317–19, 440
   thermodynamic model, 478

Energy bar charts, 249–50, 295–96, 301
Energy density, 855, 858, 988
Energy diagrams, 261–65, 813, 1140
   equilibrium positions, 263
   molecular bonds and, 264–65
Energy equation, 295, 301, 471
Energy-level diagrams, 1140–41
   for hydrogen atom, 1146–47, 1219
   for multielectron atoms, 1226, 1232
Energy levels, 1137
   allowed, 1184, 1187, 1189
   of hydrogen atom, 1143–45, 1219–20
Energy reservoirs, 527–28
Energy transfer, 246, 279–80, 470–71
   heat and energy, 471
   rate of (power), 297–98
Energy transformations, 246, 279–80, 470
Engines. *See also* Heat engines
   Carnot engine, 542–46
   diesel, 536
   gas turbine engines, 536
   gasoline internal combustion, 536
English units, 24–25
Entropy, 518–520, 557
Environment, 169, 279
Equation of continuity, 425, 429–30
Equation of motion, 387–89, 393, 395
Equilibrium, 133, 138
   dynamic, 128, 141
   electrostatic, 799–802, 848–50, 870–71
   hydrostatic, 414, 416
   mechanical, 128
   phase, 452
   problem-solving strategies, 139, 159
   static, 128, 140
   thermal, 514–20
Equipartition theorem, 510–11
Equipotential surfaces, 823, 825, 826,
   828, 840–41, 845–47
Equivalence, principle of, 358–59
Equivalent capacitance, 851, 852
Equivalent resistance, 899, 903, 904, 905
Escape speed, 363–64, 815
Estimates, order-of-magnitude, 27
Evaporation, 557
Events, 1069–70, 1073–76, 1081–83
   and observations, 1070–71
   time of, 1071
Excitation, 1232
   collisional, 1233–34
Excited states, 1231–35
   lifetime of, 1236–38
Explosions, 220, 234–36
Exponential decay, 396, 992
External forces, 169, 228, 294
Eyepiece, 704, 706
Eyes. *See* Vision

F

Fahrenheit scale, 449
Far point (FP) of eye, 701
Farad (F), 850
Faraday's law, 963, 975–81, 1015
   electromagnetic fields and, 1009–11
   and electromagnetic waves, 982,
      1017–19
   for inductors, 985–86
Farsightedness, 701
Fermat's principle, 693
Ferromagnetism, 951
Fiber optics, 665–66, 685–86
Fictitious forces, 201–04
Field diagrams, 739–40
Field model, 736–42
Fields, 737. *See also* Electric fields;
   Electromagnetic fields; Magnetic
   fields
Finite potential wells, 1193–98
First law of thermodynamics, 478–80,
   490–91, 527, 556, 1015
Fission, 1094
Flat-earth approximation, 145
Flow tube, 424
Fluids, 311, 407–40
   Archimedes' principle, 420
   Bernoulli's equation, 426–30
   buoyancy, 419–23
   density, 408–09
   displaced, 420
   dynamics, 423
   equation of continuity, 424–26
Fluorescence, 1106–07
Flux. *See* Electric flux; Magnetic flux
f-number, 698–99
Focal length, 671, 682, 684
   angular magnification and, 704
   effective focal length of combination
      lenses, 697
Focal point, 671, 674, 676, 682
Force, 1, 116–137
   acceleration and, 124–25, 141, 157–58,
      205
   action/reaction pair, 170, 172–73, 175
   buoyant, 419–23
   combining, 118
   conservative and nonconservative,
      288–89, 811
   contact, 117
   dissipative, 293–94
   drag, 121, 152–54
   electric, 121–22, 733–36
   external, 169
   fictitious, 201–04
   friction, 120–21, 148–52

   gravitational, 119, 357–58
   identifying, 122–23, 130, 133
   impulsive, 221
   as interactions, 127, 173
   internal, 234
   long-range, 117, 736–37
   magnetic, 121–22
   net force, 80, 118
   normal, 120
   restoring forces, 255–57
   SI unit, 126
   spring, 119
   superposition of, 118
   tension, 119–20, 156, 177–81
   thrust, 121
Free-body diagrams, 130–32, 133
Free fall, 51–54, 61
   weightlessness and, 147–48
Free-fall acceleration, 52
Freezing point, 449, 451
Frequency, 378–79, 382
   angular, 379, 381, 388, 594
   beat, 617
   crossover, 1039
   cyclotron, 942
   fundamental, 597, 600–02
   of mass on spring, 385
   natural, 398
   of pendulum, 392, 394
   resonance, 398
   of sinusoidal waves, 566
   of sound waves, 574–75
Friction, 120–21, 128, 148–52
   causes of, 152
   coefficients of, 148–49, 206
   kinetic, 148, 176
   model of, 149
   rolling, 149
   static, 148–49, 176
Fringe field, 765
Fringe spacing, 631–32
Fundamental frequency, 597, 600–02
Fundamental unit of charge, 725, 1111–12
Fusion, 836

G

Galilean field transformation equations,
   1007–09
Galilean relativity, 1008–09, 1061–65
   principle of relativity, 1064
Galilean transformations
   of acceleration, 1005
   of electromagnetic fields, 1007–10
   of position, 1063
   of velocity, 97, 267–68, 1004, 1063,
      1068

Gamma decay, 1266–67
Gamma rays, 1093–94, 1198, 1249, 1258–59, 1266–67
    medical uses of, 1269
Gases, 408, 445, 446
    ideal gases, 452–56
    ideal-gas processes, 456–61, 478–80
    monatomic and diatomic, 447
    pressure, 411, 505–07
    specific heats of, 485–91
Gas turbine engines, 536–38
Gauge pressure, 415
Gaussian surfaces, 784
    calculating electric flux, 785–87
    electric field and, 785
    symmetry of, 784, 791
Gauss's law, 791–809, 934, 1010, 1015, 1017
    and conductors in electrostatic equilibrium, 799–802
    Coulomb's law vs. , 791, 794
    for magnetic fields, 936
Geiger counter, 859,1259
Generators, 842, 982–83, 1034
Geomagnetism, 923, 931
Geosynchronous orbits, 367
Global positioning systems (GPS), 212
Global warming, 494
Grand unified theory, 217
Gravitational constant (G), 357, 359–61
Gravitational field, 737, 811
Gravitational force, 119, 145–46, 357, 359–61
    and weight, 357–58
Gravitational mass, 358
Gravitational potential energy, 246, 248–54, 362–65
    flat-earth approximation, 364
    zero of, 363
Gravitational torque, 324–25, 394
Gravity, 145–46, 311, 736–37, 811–12.
    See also Newton's law of gravity
    little g (gravitational force) and big G (gravitational constant), 359–61
    moon's orbit and, 201
    Newton's law of, 145
    on rotating earth, 202–03
    universal force, 356
Gray (Gy), 1268
Greenhouse effect, 494
Grounded circuits, 729, 908–09
Ground state, 1218, 1219, 1220, 1231

H

Half-life, 397, 1077, 1260
Hall effect, 944

Harmonics, 597, 602, 603
Hearing, threshold of, 579
Heat, 279, 293, 471, 475–77
    defined, 476
    in ideal-gas processes, 488
    specific heat and temperature change, 480–81
    temperature and thermal energy vs., 477
    thermal interactions, 476
    transfer mechanisms, 491–94
    units of, 477
    work and, 471, 476, 490, 527–29
Heat engines, 526–555
    Brayton cycle, 536–38
    Carnot cycle, 542–46
    ideal-gas, 534–36
    perfect, 530, 533
    perfectly reversible, 541–42
    problem-solving strategy for, 535–36
    thermal efficiency of, 529–30
Heat exchanger, 537
Heat of fusion, 482
Heat pump, 551
Heat-transfer mechanisms, 491–94
Heat of transformation, 482–83
Heat of vaporization, 482
Heisenberg uncertainty principle, 1169–72, 1190
Helium-neon laser, 1241–42
Henry (H), 984
Hertz (Hz), 378, 382
History graphs, 564–65, 566
Holography, 645–46
Hooke's law, 256–57, 289, 361, 387
Hot reservoir, 527
Huygens' principle, 637–39
Hydraulic lifts, 418–19
Hydrogen atom
    angular momentum, 1217–19
    Bohr's analysis of, 1141–46
    energy levels of, 1219
    spectrum, 1146–49
    wave functions and probabilities, 1220–23
Hydrogen-like ions, 1147–48
Hydrostatics, 413–15
Hyperopia, 701, 702

I

Ideal battery, 843, 902
Ideal-fluid model, 423
Ideal gases, 452–56
    ideal-gas heat engines, 534–38
Ideal-gas processes, 456–61, 478–80, 534
    adiabatic process, 479, 488–91, 536–39, 543–44

constant-pressure process, 458–58
constant-temperature process, 459–60
constant-volume process, 457–59
pV diagram, 456
quasi-static processes, 456–57
work in, 471–75
Ideal wire, 883
Image distance, 660, 667, 676
Image formation
    by refraction, 666–67
    with spherical mirrors, 682–85
    with thin-lenses, 680–81
Image plane, 672
Impedance, 1043
Impulse, 220–26
Impulse approximation, 225
Impulse-momentum theorem, 222–23, 226
    similarity to work-kinetic energy theorem, 281–82
Impulsive force, 221
Inclined plane, motion on, 54–58
Independent particle approximation (IPA), 1226, 1227
Index of refraction, 576–77, 609, 646, 662, 663, 664
Induced current, 963–78, 985, 986
    in a circuit, 966–67
    eddy currents, 968
    Faraday's law, 975–78, 1009
    Lenz's law, 971–74
    magnetic flux and, 968–71
    motional emf, 964–65
Induced electric dipole, 754–55
Induced electric fields, 856, 978–981, 1010, 1013
Induced emf, 975–78, 985, 986
Induced magnetic dipoles, 951–53
Induced magnetic fields, 981–82, 985, 1013–14
Inductance, 984, 985
Inductive reactance, 1041–42
Inductor circuits, 1041–42
Inductors, 984–88
Inelastic collisions, 232–34
Inertia, 126
    law of, 128
Inertial mass, 126, 318, 358–59
Inertial reference frames, 129, 201, 202, 1064
Insulators, 724, 727–31, 883
    dielectrics, 856–58
Intensity, 578–80, 1025
    of double-slit interference pattern, 633–34
    of electromagnetic waves, 1021–22
    of standing waves, 594

Interacting systems, 168–72
  analyzing, 169–70
  revised problem-solving strategy for,
    175–77
Interaction diagrams, 169, 170
Interference, 594, 604–610. *See also*
    Constructive interference; Destructive
    interference
  of light, 596, 629–34
  in one dimensional waves, 604–07
  mathematics of, 607–10
  and phase difference, 605–07
  photon analysis of, 1159–60
  problem-solving strategy for, 613
  in two- and three-dimensional waves,
    610–15
  wave analysis of, 1157–58
Interference fringes, 630, 631–32, 644
Interferometers, 642–46, 1135–36
  acoustical, 643
  atom, 1135–36
  Michelson, 644–45
Internal energy, 470
Internal resistance, 901–02, 905
Inverse-square law, 354, 356, 357
Inverted images, 672
Ion cores, 727
Ionization, 726, 1108
Ionization energy, 1145, 1220, 1231
Ionization limit, 1146, 1147
Ionizing radiation, 1259
Ions, 726
  hydrogen-like ions, 1147–48
Irreversible processes, 516–17
Isobaric (constant-pressure) processes,
    458–59, 474, 488
Isobars, 1250
Isochoric (constant-volume) processes,
    457, 474, 479, 488
Isolated systems, 220, 228, 231–32, 234
  conservation of energy, 295
  conservation of mechanical energy,
    254–55
  second law of thermodynamics, 519–20
Isothermal (constant-temperature)
    processes, 459–61, 474–75, 479
Isotherms, 460, 485
Isotopes, 1118, 1250

J

Joules (J), 248, 281, 454

K

Kelvin scale, 450, 452
Kelvins (K), 451

Kepler's laws of planetary motion, 354,
    355, 365–68
Kinematics, 1, 33–68
  circular motion, 98–107
  with constant acceleration, 45–51
  with instantaneous acceleration, 58–60
  free fall, 51–54
  in two dimensions, 87–91, 108
  uniform motion, 34–38
Kinetic energy, 246, 247–51
  in elastic collisions, 265–69
  in relativity, 1090–91
  of rolling object, 335–36
  rotational, 312, 317–19
  temperature and, 508–09
  work and, 280–82
Kinetic friction, 120, 148–49, 155, 206
Kirchhoff's laws,
  junction law, 878, 892–93
  loop law, 847, 893–95, 989, 991, 1035

L

Laminar flow, 423
Lasers
  creating using stimulated emission,
    1238–42
  helium-neon laser, 1241–42
  quantum-well laser, 1197
  ruby laser, 1240–41
Lateral magnification, 674, 684, 698, 704
LC circuits, 988–91
Length contraction, 1078–82, 1085
Lenses, 670. *See also* Cameras; Thin
    lenses
  aberrations, 707–10
  achromatic doublet, 715
  angular resolution, 709
  in combination, 695–97
  converging, 671
  diffraction limited, 708–09
  diverging, 695, 697, 701
  f-number of, 698–99
  focal length, 695–98
  ray tracing, 670–73, 695–96
Lens maker's equation, 680
Lens plane, 671
Lenz's law, 971–74, 985
Lever arm, 323–25, 394
Lifetime (of excited states), 1236–38
Lift, 429–30
Light, 627–654, 982. *See also*
    Electromagnetic waves
  absorption or reflection by objects,
    1235
  color and dispersion, 667–70
  early theories of, 1103

emission and absorption of, 1103–06
  interference of, 629–34
  models of, 628–29
  photon model of, 629, 1133–34
  properties of, 1278
  ray model of, 629, 641–42, 656–58
  wave model of, 629, 641–42
Light clock, 1074, 1075
Light rays, 656
Light waves, 575–77. *See also*
    Electromagnetic waves
  Doppler effect for, 582–83
  interference of, 604, 608–10
  polarization of, 1024–26
Light years, 1077
Line of action, 323
Linear acceleration, 12–13
Linear charge density, 757, 759
Linear density, 562, 570
Linear restoring force, 393
Line of charge, 758–60
Line integrals, 934–935, 936, 937, 939
Line of nuclear stability, 1253
Liquid-drop model, 1252
Liquids, 408, 445. *See also* Fluids
  pressure in, 413
Longitudinal waves, 561, 599
Long-range forces, 117, 721, 736
Lorentz force law, 1015
Lorentz transformations, 1082–87, 1090
Loschmidt number, 522
LC circuits, 988–91
LR circuits, 991–93
Lyman series, 1147

M

Macrophysics, 292
Macroscopic systems, 444–46, 470
Magnetic dipole moment, 933, 949
Magnetic dipoles, 922, 931–34
  induced, 951–53
Magnetic domains, 951, 952
Magnetic field lines, 924
Magnetic field strength, 925, 988
Magnetic fields, 923–39, 1009
  Ampère's law, 934–37
  Biot-Savart law, 925, 926, 927, 928
  of current, 927–31
  of current loop, 930–31
  of cyclotron, 944
  energy in, 984, 988
  Gauss's law for, 1010
  induced, 981–82, 1013–14
  of moving charge, 925–27
  properties of, 924
  of solenoids, 938–39

Magnetic fields (*continued*)
transformations of, 1005–10
uniform, 038
Magnetic flux, 962, 968–970, 982, 983, 984, 985, 994
Faraday's law, 975–78
Lenz's law, 972
in nonuniform field, 970–71
Magnetic force, 121–22
on current-carrying wires, 924–25, 932
on moving charge, 925–26
Magnetic poles, 922, 932, 948
Magnetic quantum number, 1218
Magnetic resonance imaging (MRI), 939, 952, 1269–70
Magnetism, 719, 922, 1056
ferromagnetism, 951
magnetic properties of matter, 950–53
Magnification, 704–07
angular, 704
lateral, 674, 684, 698
Magnifier/magnifying glasses, 675, 703
Malus's law, 1024–25
Manometers, 416
Mass, 126, 133, 138, 144, 1057
atomic, 447
conservation of, 219, 1093
equivalence to energy in relativity, 1092–94, 1250
gravitational, 358–59
inertial, 126, 318
measurement of, 24
molar, 448
molecular, 447
weight, gravitational force, and, 119, 144–147
Mass density, 408–09, 445
Mass-energy equivalence, 1092–94, 1250
Massless string approximation, 178–80
Mass number, 1118, 1249
Mass spectrometers, 960, 1117
Matter waves, 1134–36
Maxwell's equations, 1014–16, 1016, 1066
Mean free path, 503–04
Mean time between collisions, 874
Mechanical energy, 254
conservation of, 254–55, 318
conservative forces and, 289–90
Mechanical equilibrium, 128, 471
Mechanical interaction, 471
Mechanical waves, 561
Medium, 561
displacement of particles by wave, 566
electromagnetic waves and, 661
speed of sound in, 574
wave speed in, 563

Melting point, 445, 449, 482–83
Metal detectors, 984
Metals, 868–69, 879, 882–83, 1056
Meter (m), 23, 1208
Michelson interferometer, 644–45
Micro/macro connection, 502–525
equilibrium, 517–18
gas pressure, 505–07
irreversible processes, 516–17
molecular collisions, 503–04
order, disorder, and entropy, 518–19
second law of thermodynamics, 519–20
temperature and, 508–09
thermal interactions and heat, 514–16
Microphysics, 292
Microscopes, 675, 704–05, 710, 711
Millikan-oil-drop experiment, 1111–12
Minimum spot size, 708–09
Mirror equation, 684–85
Mirrors
plane mirrors, 659–60
spherical mirrors, 682–85
Models, 1, 21
atomic model, 120, 221, 264
basic energy model, 245, 246–47, 279–80, 301
Bohr's model of the atom, 1138, 1141–46
charge escalator model, 843
charge model, 721–25
of electrical conduction, 873–74
electric field models, 751
field model, 736–42
of friction, 149
ideal-fluid model, 423
ideal-gas model, 452–56
nuclear model of atoms, 1114, 1116–17
particle model, 4–5
photon model of light, 629, 1133–34
quantum-mechanical models, 1138–41, 1148
raisin-cake model of atoms, 1112–13
ray model of light, 629, 641–42, 656–58
rigid-body model, 311, 313
shell model of atoms, 1222, 1228–31, 1256–58
thermodynamic energy model, 478
wave model of light, 561–63, 629, 641–42
Modes, 597, 619
Molar mass, 447–48
Molar specific heats, 481, 510–13
at constant pressure, 486
at constant volume, 486
Molecular bonds, 264–65
covalent, 1205–06

Molecular mass, 447
Molecular vibrations, 265, 1202–03
Moles, 447–48
Moment arm, 323–25, 394
Moment of inertia, 311, 317–21, 335
Momentum, 222
angular, 340–45
changes in kinetic energy and, 282
conservation of, 226–32, 1067
and impulse, 221–23
problem-solving strategy for, 223–26
quantization of angular momentum, 1145–46
relativistic, 1087–90
in two dimensions, 236–37
velocity-energy-momentum triangle, 1092
Momentum bar charts, 223
Monatomic gases, 447, 453, 487
Motion, 1–32, 216. *See also* Acceleration; Circular motion; Kinematics; Linear motion; Newton's laws of motion; Oscillations; Projectile motion; Relative motion; Rotational motion; Simple harmonic motion (SHM); Uniform circular motion; Velocity
of charged particle in electric field, 767–69
with constant acceleration, 90–91, 124
cyclotron, 941–43
graphical representations of, 1, 2, 17–19, 45, 56–58, 61
on inclined plane, 54–58
in one dimension, 15–19
types of, 3
uniform, 34–37
vectors, 2, 6
Motional emf, 964–68, 1009
Motion diagrams, 3–6, 38
acceleration vectors, 13–14
displacement vectors, 9
examples, 14–16
velocity vectors, 11–12
Motion graphs, 56–58, 61
Motors, 949, 1048–49
MRI (magnetic resonance imaging), 939, 952, 1269–70
Myopia, 702–03

N

Natural frequency, 398–99
Near point (NP) of eye, 701
Nearsightedness, 702–03
Neutral buoyancy, 421
Neutrino, 1266
Neutron number, 1250

Neutrons, 1117–18, 1197, 1249
Newtons (N), 126, 281
Newton's first law of motion, 127–29, 133, 139, 216, 390
Newton's law of gravity, 145–46, 354–76, 440
Newton's second law of motion, 126–27, 133, 141–44, 205, 207, 216
    examples of, 155–58
    for oscillations, 387, 570, 571
    for rotational motion, 326, 327, 330
    in terms of momentum, 222
Newton's third law of motion, 172–77, 216, 472, 505
    conservation of momentum and, 226–29
    problem-solving strategy for interacting objects, 175–77
    reasoning within, 173–74
Newton's zeroth law, 127
Nodal lines, 612
Nodes, 594, 595
Nonconservative forces, 289, 294
Normal force, 120, 203–04
Normalization, 1164–66, 1188
Normal modes, 597
Nuclear decay, 1263–67
Nuclear fission, 1094
Nuclear force, 1254, 1255–56
Nuclear fusion, 836
Nuclear magnetic resonance (nmr), 1270
Nuclear model of the atom, 1114, 1116–17
Nuclear physics, 1197–98, 1248–79
    biological applications, 1268–71
    decay mechanisms, 1263–67
    nuclear size and density, 1251–52
    nuclear stability, 1252–1255
    nucleons, 1197, 1249
    properties of nuclei, 1278
    shell model, 1256–57
    strong force, 1255–56
Nucleons, 1197, 1249
Nucleus, 1114. See also Nuclear physics
    discovery of, 1112–17
    nuclear size and density, 1251–52
Number density, 446

O

Object distance, 667, 676, 679, 682
Objective, 704
Object plane, 672
Ohm (Ω), 880
Ohmic materials, 882, 883
Ohm's law, 881, 882–84, 1035
One-dimensional waves, 564–66

Optical axis, 667
Optical cavity, 1240
Optical instruments, 694–715
    for magnification, 703–07
    resolution of, 707–10
Optics, 628, 716. See also Light; Optical instruments; Ray optics; Wave optics
Orbital angular momentum, 1218
Orbital quantum number, 1217
Orbits, 354
    circular, 199–201, 355, 371
    elliptical, 355
    energetics of, 369–70
    geosynchronous, 367
    Kepler's laws, 355, 365–68
Order (of diffraction), 635
Oscillations, 257, 262–63, 311, 377–406, 440. See also Simple harmonic motion (SHM)
    amplitude of, 378
    angular frequency, 379
    damped, 395–99
    driven, resonance and, 398–99
    frequency of, 378
    initial conditions, 381–84
    period of, 378, 562
    phase of, 382
    turning points in, 363, 379
Oscillators, 378, 388–89
    quantum harmonic oscillator, 1200–05
Otto cycle, 536

P

Parallel-axis theorem, 321, 346
Parallel-plate capacitors, 764–66, 787, 849–59
    electric field of, 765–66, 841
    electric flux inside, 787–88
    electric potential of, 821–25, 841
Paraxial rays, 667, 676–77
Parent nucleus, 1263–65
Particle accelerators, 943
"Particle in a box"
    energies and wave function, 1183–88
    interpreting the solution, 1188–91
    potential energy function for, 1185, 1189
Particle model, 4–5, 117
Particles, 4
Pascal (Pa), 410, 417
Pascal's principle, 414
Path-length difference, 605–06, 611, 612, 630–31, 643
Pauli exclusion principle 1227–28, 1257
Pendulums, 268–69, 311, 391–94
    ballistic, 253–54, 344–45

damped, 397
physical, 394
Penetration distance, 1196
Perfect destructive interference, 605, 606, 608, 611, 612, 619
Perfectly elastic collisions, 265
Perfectly inelastic collisions, 232
Perfectly reversible engine, 540–42
Period, 98, 366
    of oscillation, 378, 392, 562
    of planetary bodies, 367
    of sinusoidal waves, 566
Periodic table of the elements, 1228–31, A-4
Permanent magnets, 932, 952
Permeability constant, 925
Permittivity constant, 733
Phase (oscillation), 382
Phase (wave), 573–74
Phase angle, 1044
Phase changes of matter, 445, 450–52, 482–83
Phase constant, 382–84, 569
Phase diagram, 451
Phase difference, 573, 605–07, 611
Phase equilibrium, 451
Phasors, 1024
Photodissociation, 265
Photoelectric effect, 1126, 1209
    classical interpretation of, 1127–29
    Einstein's explanation of, 1129–32
Photoelectrons, 1126
Photolithography, 708
Photon model of light, 629, 1133–34
Photons, 628, 1067, 1093, 1132–34. See also Light
    absorption and emission, 1103–06
    connecting wave and photon views of interference, 1160–61
    energy of, 1133–34
    photon emission rate, 1134
    photon model of light, 629, 1133–34
Photosynthesis, 669–70
Physical pendulum, 394
Pictorial representations, 19–21
Pinhole cameras, 657
Pivot point, 322, 330, 394
Planck's constant, 1093, 1129, 1135
Plane mirror, 659–60
Plane of polarization, 1024
Plane waves, 572, 1016, 1017, 1021
Planets. See also Orbits
    extrasolar, 368
    Kepler's laws of planetary orbits, 355, 365–68
Plasma ball, 827
Point charges, 732, 733

Point charges (*continued*)
  electric field of, 739–40, 751
  electric field of multiple point charges, 752–56
  magnetic field of, 927
  potential energy of, 814–17
Point source of light rays, 657
Polarization
  charge polarization, 729–31
  of electromagnetic wave, 1024–26
  Malus's law, 1025
Polarization force, 265, 729–30, 770
Polarizing filters, 1024
Polaroid, 1024
Poles, magnetic, 922, 932, 948
Population inversion, 1240
Position vectors, 6–7
Position-versus-time graphs, 17–19, 38–39, 45
Positron-emission tomography (PET scans), 1094
Positrons, 1093, 1264
Potassium-argon dating, 1262
Potential differences, 820, 822, 944–45, 964–65
  across batteries, 843–44
  across capacitors, 849–53
  across inductors, 985–87
  across resistors, 894, 895
Potential energy, 246, 295
  conservative force and, 288–89, 294, 811
  elastic, 257–61
  electric, 811–18
  finding force from, 290–92
  gravitational, 246, 248–54, 362–65
  in mechanical energy, 254–55
  at microscopic level, 292–93
  inside the nucleus, 1198
  work and, 288–90
  zero of, 824, 908
Potential-energy curve, 261–63
Potential-energy function, 1182, 1183, 1184, 1189
Potential wells, 1193–1198, 1205
  classically forbidden region, 1195–96
  nuclear physics, 1197–98
Power, 297–300
  in AC circuits, 1046–49
  in DC circuits, 896–98
  of lenses, 701
  of waves, 578–80
Power factor, 1048–49
Poynting vector, 1021
Prefixes, denoting powers of ten, 24
Presbyopia, 701
Pressure, 409–15
  atmospheric, 411–13

blood pressure, 417–18
  causes of, 410–11
  constant-pressure process, ideal gases, 458–59
  in gases, 411, 453–55, 505–07
  in liquids, 413–15
  measuring, 415
  units of, 417
Pressure gauge, 407, 415
Principal quantum number, 1217
Prisms, 668
Probabilities, 1158
  of detecting particle, 1162–64
  of detecting photon, 1160, 1162–63, 1190–91
Probability density, 1161, 1164
  corresponding classical quantity, 1192–93
  radial probability density, 1221
Projectile motion, 3, 91–95, 193
  launch angle, 92
  problem-solving strategy for, 94–95
  reasoning about, 93
Proper length, 1079–80
Proper time, 1075–76
Proportionality, 125
Proportionality constant, 124, 125
Propulsion, 171–72
Protons, 1116, 1197, 1249
Pulleys, 175, 180
pV diagrams, 456

Q

Quadrants of coordinate system, 74
Quanta of light, 1130–32
Quantization, 1129, 1184
  of angular momentum, 1145–46, 1218–19
  of charge, 726
  Bohr's model of atomic quantization, 1138–41
  of energy, 1136–38
Quantum computers, 1279
Quantum harmonic oscillator, 1200–05
Quantum jumps, 1184, 1189, 1198
Quantum-mechanical models, 1182
Quantum mechanics, 1137,
  correspondence principle, 1191–93
  drawing wave functions, 1199
  law of, Schrödinger equation, 1180–1184
  particle in a box, energies and wave function, 1183–88
  particle in a box, interpreting solution, 1188–91
  potential wells, 1193–1198

problem-solving strategy, 1184, 1217
  wave functions, 1162
Quantum numbers, 1137, 1225
  in hydrogen atoms, 1217–18
  and Pauli exclusion principle 1227–28
  in protons and neutrons, 1257
Quantum-well laser, 1197
Quasars, 583
Quasi-static processes, ideal gases, 456–57, 473

R

Radial acceleration, 105
Radial axis, 194, 195
Radial probability density, 1221
Radial wave functions, 1221–23
Radians, 99, 379
Radiated power, 493
Radiation, 1258–63
  blackbody radiation, 493, 1104
  medical uses of radiation, 1269
  radiation dose, 1268–69
  radioactivity, 1113, 1249
  thermal radiation, 493–94
Radiation pressure, 1022–23
Radio waves, 982
Rainbows, 669
Rarefactions, 574, 599
Rate equations, 1237–38
Ray diagrams, 657
Ray optics, 655–93
  color and dispersion, 667–70
  ray model of light, 656–58
  reflection, 658–60
  refraction, 661–70
Ray tracing, 670–76, 682–83
Rayleigh scattering, 670
Rayleigh's criterion, 709
RC circuits, 909–12
RC filter circuits, 1038–41
Real images, 672, 675, 677, 680, 682
Red shift, 582
Reference frames, 96–98, 1061–62, 1064
  accelerating, 129
  in Einstein's principle of relativity, 1066, 1078
  in Galilean relativity, 1061–62
  inertial, 129
Reflection, 658–60
  diffuse, 659
  law of, 658, 659
  specular, 658
  total internal reflection (TIR), 664–65
Reflection gratings, 636
Refraction, 661–66
  image formation by, 666–67

index of refraction, 576–77, 609, 646, 662

sign conventions for refractive surfaces, 677

Snell's law of, 661

total internal reflection (TIR), 664–65

Refrigerators, 526, 532–33

coefficient of performance, 532–33

ideal-gas, 538–39

perfect, 533

Relative biological effectiveness (RBE), 1268

Relative motion, 95–98, 108

Relativity, 441, 1008–09, 1278. *See also* Galilean relativity

causal influence, 1089–90

clock synchronization, 1070

Einstein's principle of, 1066–68

energy and, 1090–95

events, 1068–69, 1070–71

Galilean, 1008–09, 1061–65

general, 1061

length contraction, 1078–82

Lorentz transformations, 1082–87

measurements, 1069–70

momentum and, 1087–90

proper time, 1075

simultaneity and, 1071–75

special, 1061

time dilation, 1074–78

Resistance, 880–84

equivalent, 899, 903

internal, 901, 905

Resistivity, 879

Resistor circuits, 906–08, 1034–36

Resistors, 882, 883

Ohm's law and, 893, 898

parallel resistors circuit, 903–05

power dissipated by, 897

series resistors circuit, 1042–46

Resolution,

angular, 709

of diffraction grating, 653

of optical instruments, 707–09

Resonance, 398

LC circuits, 1044–45

mechanical resonance, 398–99

standing-wave resonance, 603

Resonance frequency, 398, 1044

Rest energy, 1091

Rest frame, 1074

Restoring forces, 255–57, 361, 387

Resultant vector, 71

Right-hand rule, 338, 923, 927, 946

Rigid bodies, 312. *See also* Rotational motion

Rigid-body model, 313

Ring of charge

electric field of, 760–61, 844, 872

electric potential of, 829–30

RLC circuits, series, 1042–46

Rocket propulsion, 171, 236

Rolling constraint, 334

Rolling friction, 149, 153

Rolling motion, 334–37

Root-mean-square (rms) current, 1046–47

Root-mean-square speed (rms speed), 506–07

Ropes and pulleys, 177–81

acceleration constraints, 175

massless string approximation, 178–80

tension, 177–78

Rotational kinematics, 98–107, 313–14, 346

Rotational kinetic energy, 312, 317, 508, 511

Rotational motion, 3, 103–05, 311, 313–14, 440

angular momentum, 340–45

about the center of mass, 314–17

about a fixed axis, 327–29

dynamics, 325–27

rolling motion, 334–37

torque, 312, 321–25

vector description of, 337–40

rtz coordinate system, 193–94, 209

Rutherford model of atom, 1114

S

Satellites, 365

orbital energetics, 369–70

orbits, 365–68

s-axis, 35–36

Scalar product, 337–40

Scalars, 6, 70

Scanning tunneling microscope (STM), 1209

Schrödinger equation, 1180–82, 1228

solving, 1183–84

Screening, 800

Sea of electrons, 727

Second law of thermodynamics, 519–20, 527, 533, 541, 545, 556, 1017

Selection rules, 1232

Self-inductance, 984

Series *RLC* circuits 1042–46

Shell model of atom, 1222, 1228–31, 1256–57

Short-range forces, 217, 1197

Sieverts (Sv), 1268

Sign convention

for electric and magnetic flux, 1018

for motion in one dimension, 16–17

for refracting surfaces, 677

for rotational motion, 314

for spherical mirrors, 684

for thin lenses, 680

Significant figures, 2, 25–27

SI units, 2, 23–25

Simple harmonic motion (SHM), 378–81, 569, 595, 1034

and circular motion, 381–84

dynamics of, 386–89

energy in, 384–86

kinematics of, 379–81

Simultaneity, 1071–75, 1089

Single-slit diffraction, 636–40

Sinusoidal waves, 366–72

fundamental relationship for, 567–68

mathematics of, 568–70

standing waves, 593

wave motion on a string, 570–72

Small-angle approximation, 391–93

Snapshot graphs, 563–64, 566

Snell's law, 661, 662, 663, 667, 677

Sodium

emission spectra, 1234

excited states of, 1232

Solenoids, 938–39, 952. *See also* Inductors

Solids, 445–46

color in, 1234–35

induced electric field in, 980

phase changes, 450–52, 482–83

specific heat of, 511–12

Sound intensity levels, 579–80

Sound waves, 561, 574–75

beats, 615–18

Doppler effect, 580–83

standing sound waves and musical acoustics, 599–604

Source charges, 737, 738, 740, 821

Spacetime coordinates, 1069, 1081–1083

Spacetime interval, 1081–82

Special relativity, 1061

Specific heat, 480–81

of gases, 485–91

thermal energy and, 510–13

Specific heat ratio, 489

Spectrometer, 1103

Spectroscopy, 636

Spectrum, 1103. *See also* Absorption spectrum; Emission spectrum

excited states and spectra, 1231–35

hydrogen atom spectrum, 1146–49

Specular reflection, 658

Speed, 35

escape, 363–64, 815

of light, 575, 1020, 1066–67

molecular, 503–04

Speed (*continued*)
root-mean-square (rms), 1046–47
of sound, 574
terminal, 154
velocity vs., 11
wave, 561–63
Sphere of charge, 764, 795–97
Spherical aberration, 707
Spherical mirrors, 682–86
Spherical symmetry, 783, 784
Spherical waves, 572, 610–15
Spin, of electrons, 950–51, 952, 1219
Spin quantum number, 1224
Spontaneous emission, 1238
Spring constant, 255–56
Spring force, 119, 124
Springs. *See also* Oscillations; Simple
harmonic motion (SHM)
elastic potential energy, 257–60
restoring forces and Hooke's law,
255–57, 289
work-kinetic theorem for, 287
Spring scale, 272
Stability and balance, 332–34
Stable equilibrium, 263
Stable isotopes, 1250
Standard atmosphere (atm), 413, 417
Standing waves, 591, 593–603, 1136,
1142, 1190. *See also* Superpositon
electromagnetic waves, 598–99
mathematics of, 594–95
nodes and antinodes, 594
sound waves and musical acoustics,
599–604
on a string, 595–98
State variables, 445–46, 453–54
Static equilibrium, 128, 140. 330–34
Static friction, 121, 148, 157
Stationary states, 1139, 1140, 1141, 1184
allowed energies for, 1187–88
hydrogen atom, 1142, 1143, 1144,
1217–18
Stern-Gerlach experiment, 1223–23
Stick-slip motion, 256–57
Stimulated emission, 1238–42
Stopping potential, 1126, 1127–28, 1131–32
STP (standard temperature and pressure),
455
Strain, 431–32
Streamlines, 424
Stress, 417, 430–32
Strong force, 217, 1197, 1255–56
Subatomic particles, 1111
Sublimation, 451
Superconductivity, 879–80, 933
Superposition, 591–626, 751, 828
beats, 615–18

creating a wave packet, 1167
of electric fields, 751–52, 757, 759
of forces, 118
of magnetic fields, 926–27
principle of, 592–93
of two or more quantum states, 1279
Surface charge density, 757, 821
Surface integrals, 785–86, 788–89
Symmetry
of electric fields, 781–83, 791
of magnetic fields, 930, 931, 934
Systems, 169, 219, 227
disordered, 518–19
energy of, 279–80, 293, 295
isolated, 220, 228
ordered, 518–19
self-organizing, 557
total momentum of, 227–30, 260

T

Tangential acceleration, 105–06, 205
Tangential axis, 194
Telescopes, 706–07, 711
resolution of, 709
Temperature, 449–54, 508–09
absolute, 450
change in, and specific heat, 480
heat and thermal energy vs., 477
Tensile strength, 431–32
Tensile stress, 430–32
Tension force, 119–20, 156
Terminal speed, 154
Tesla (T), 925
Thermal conductivity, 492
Thermal efficiency, 529–30, 537–38
limits of, 540–42
Thermal energy, 246, 247, 254, 279, 292–94
heat and temperature vs., 449, 477
in inelastic collisions, 265
properties of matter, 480–83
and specific heat, 510–13
Thermal equilibrium, 446, 453–54, 476,
514–20
Thermal interactions, 471, 476
Thermal properties of matter, 480–83
Thermal radiation, 1104
Thermodynamic energy model, 478
Thermodynamics, 443–468, 527, 556
first law of, 480–82, 490–91, 556, 1015
nonequilibrium, 557
second law of, 519–20, 527, 533, 541,
545, 556, 1015
Thermometers, 497
Thin-film optical coatings, 608–10
Thin lenses, 670–76
ray tracing, 670–76, 682–83

refraction theory, 676–81
sign conventions for, 680
Threshold frequency, 1126, 1131
Thrust, 121, 171
Time
direction or arrow of, 519–20
measurement of, 10, 23
spacetime coordinates, 1069, 1081
Time constant
in *LR* circuits, 991, 992
in oscillations, 396, 397, 399
in *RC* circuits, 910
Time dilation, 1074–78
Torque, 312, 321–25, 346
angular acceleration and, 313–14
on current loops, 948–50
gravitational, 324–25
net torque, 324, 341
torque vector, 339–40
Total internal reflection (TIR), 664–66
Total momentum, 227–28
Trajectory, 3, 87–89, 141. *See also*
Projectile motion
parabolic, 199, 204
in projectile motion, 193, 216
Transformers, 983–84
Transitions, 1220, 1232
nonradiative, 1235
radiative, 1239
Translational kinetic energy, 508–15
Translational motion, 3, 313
Transmission grating, 636
Transverse waves, 561
Traveling waves, 558–590
amplitude of, 566
displacement, 566
Doppler effect, 580–83
electromagnetic waves, 575–76
frequency of, 566
power, intensity, and decibels, 578–80
sinusoidal waves, 566–72
spherical waves, 572
types of, 561
Triple point, 452
Tsunami, 717
Tunneling current, 1209
Turbulent flow, 423, 424
Turning points, 41, 43, 261, 262, 263, 379
Twin paradox, 1077–78

U

Ultrasonic frequencies, 575
Uncertainty, 1168–69
Uncertainty principle, 1169–72
Uniform circular motion, 79, 98–103,
108, 193–99, 216

acceleration in, 86–87, 101–03
dynamics of, 195–99
simple harmonic motion (SHM) and, 381–84
velocity in, 101–03, 194–95
Uniform electric fields, 766, 770, 811–14
electric potential energy of charge in, 812–14
Uniform magnetic fields, 938
Uniform motion, 34–38
Unit vectors, 69, 77, 740, 927
Units, 2, 23–27
Universal constant, 361
Universal gas constant, 454
Unstable equilibrium, 263
Upright images, 675

V

Vacuum, 411–13
Van Allen radiation belt, 943
Van de Graaf generators, 842
Vapor pressure, 411, 416
Vector algebra, 77–80
addition, 7, 71–73, 77–78
multiplication, 73, 79, 927
subtraction, 8, 73, 79
Vector product, 338–39
Vectors, 2, 6, 69–84
area vector, 787, 970
components, 74–77
displacement, 6–9
magnitude and direction, 70
notation, 6
properties of, 70–74
unit vectors, 69, 77, 740, 927
zero, 8
Velocity, 10–12, 1067
angular, 99–101, 108, 205, 313–14
finding from acceleration, 58, 59
finding position from, 42–45
Galilean transformation of, 97, 267–68, 1004, 1063, 1068
instantaneous, 38–42
Lorentz transformation of, 1086–87
magnetic fields and force dependent on, 1004
momentum and, 222
relative, 95–96
sign of, 16–17, 61
speed vs., 11
in uniform circular motion, 101–03, 194
Velocity vectors, 11
Velocity-versus-time graphs, 39, 40, 42, 45

Venturi tube, 429
Vibrational energy levels, 1202
Virial theorem, 1123
Virtual images, 660, 674–75, 677, 680, 683
Viscosity, 423
Visible spectrum, 576, 668
Vision, 700–03
Visualizing physics problems, 22–23
Volt (V), 819
Voltage, 820, 983
of a battery, 820
of capacitors, 849, 856–59, 1036
Hall, 944–45
of inductors, 967, 986–87
peak, 1034–35
of resistors, 1035
rms, 1046–47
terminal, 843, 901
Voltmeters, 905
Volume, 408–09, 445–46, 453–56
flow rate, 425
ideal gas processes, 449, 456–460
unit, 409
Volume strain, 432
Volume stress, 432–33. *See also* Pressure

W

Waste heat, 530
Water molecules, 770
Watts (W), 297–98
Wave fronts, 572
Wave functions, 1162–63
drawing, 1199
finding, 1186
normalizing, 1166–68, 1188
radial wave function, 1221–23
Wavelengths, 567
de Broglie wavelength, 1135
and index of refraction, 577
of light waves, 576, 632, 668
measuring, 643
of sinusoidal waves, 567–68
of sound waves, 574–75
Wave model of light, 629, 641–42, 662, 707–09
Wave number, 569
Wave optics, 627–654
circular-aperture diffraction, 640–42
diffraction grating, 634–36
interference of light, 629–34
single-slit diffraction, 636–40

Wave packets, 1166–69
photons as, 1134
uncertainty about, 1168–71
Waves, 559–590, 716. *See also* Electromagnetic waves; Light waves; Sinusoidal waves; Sound waves; Standing waves; Traveling waves
amplitude of, 566
circular, 572
displacement, 566
Doppler effect, 580–83
frequency of, 566
longitudinal, 565
matter waves, 1134–36
medium of, 561
phase of, 573–74
plane, 572–73
power, intensity, and decibels, 578–80
sinusoidal, 566–72
spherical, 572
traverse, 561
Weber (Wb), 970
Weight, 119, 138, 146–47
gravitational force and, 357–58
mass vs., 146
Weightlessness, 147–48, 200
Wien's law, 1104
Work, 278–309, 470, 811, 818
basic energy model, 279–80
calculating and using, 282–86
heat and, 471, 476, 490, 527–29
in ideal-gas processes, 471–75
and kinetic energy, 280–82
and potential energy, 288–90, 362
Work function, 1127
Work-kinetic energy theorem, 281–82, 289, 294

X

X-rays, 1108, 1269

Y

Young's double-slit experiment, 629–34, 668
Young's modulus, 431–32

Z

Zero-point motion, 1190
Zero vector, 8, 73
Zoom lenses, 697–98